Renewable Geothermal Energy Explorations

Edited by Basel I. Ismail

Published in London, United Kingdom

IntechOpen

Supporting open minds since 2005

Renewable Geothermal Energy Explorations
http://dx.doi.org/10.5772/intechopen.76275
Edited by Basel I. Ismail

Contributors
Kasumi Yasukawa, Youhei Uchida, Sheng-Rong Song, Yi-Chia Lu, Alfonso Aragón-Aguilar, Jon Gluyas, Alison Auld, Charlotte Adams, Catherine Hirst, Simon Hogg, Jonthan Craig, Basel I. I. Ismail

First published in London, United Kingdom, 2019 by IntechOpen
IntechOpen is the global imprint of INTECHOPEN LIMITED, registered in England and Wales, registration number: 11086078, The Shard, 25th floor, 32 London Bridge Street
London, SE19SG – United Kingdom
Printed in Croatia

British Library Cataloguing-in-Publication Data
A catalogue record for this book is available from the British Library

Additional hard copies can be obtained from orders@intechopen.com

Renewable Geothermal Energy Explorations
Edited by Basel I. Ismail
p. cm.
Print ISBN 978-1-78984-608-9
Online ISBN 978-1-78984-609-6

For EU product safety concerns:
IN TECH d.o.o., Prolaz Marije Krucifikse Kozulić 3, 51000 Rijeka, Croatia, info@intechopen.com or visit our website at intechopen.com.

Meet the editor

Dr. Basel I. Ismail is currently an Associate Professor and Chair of the Department of Mechanical Engineering, Lakehead University, Thunder Bay, Ontario, Canada. In 2004, Professor Ismail earned his Ph.D. degree in Mechanical Engineering from McMaster University, Hamilton, Ontario, Canada. From 2004 to 2005, Dr. Ismail worked as a postdoctoral researcher at the Engineering Physics Department, McMaster University. His specialty is in engineering heat transfer, engineering thermodynamics, energy conversion, and storage engineering. Dr. Ismail's research activities are theoretical and applied in nature. Currently, his research areas of interest are primarily focused on green engineering technologies related to alternative and renewable energy systems for power generation, heating, and cooling. This includes applications and innovations in solar energy, wind energy, hydropower, geothermal and ocean energy, thermoelectricity, and other similar areas. Dr. Ismail also has an interest in other engineering-based research areas, such as energy efficiency, optimization of thermal and power systems, energy and exergy analyses of energy systems, solar/wind-powered desalination systems, and carbon-dioxide membrane gas separation (for industrial emissions control and GHG reduction). Dr. Ismail was the leading research investigator in a collaborative project (2007-2010) with Goldcorp-Musselwhite Canada Ltd. and Engineering of Lakehead University. This innovative project was state-of-the-art in geothermal energy-related technology applied in Northwestern Ontario, Canada. Dr. Ismail has published many technical reports and articles related to his research areas in reputable international journals and conferences and participated in international forums related to renewable energy technologies. During his research activities, Professor Ismail has supervised and trained many graduate students and senior undergraduate students in mechanical engineering with projects and theses related to innovative renewable and alternative energy engineering and technologies.

Contents

Preface

The geothermal resources of the Earth are enormous. The natural heat energy from the Earth is called "Geothermal Energy". It is considered to be an environmentally friendly clean energy source that could significantly contribute to the reduction of GHG emissions when utilized for electrical power generation or direct heating applications. Geothermal energy resources vary geographically, depending on the depth and temperature of the resource, the rock chemical composition, and the abundance of ground water. The source of geothermal energy is the continuous heat energy flux flowing from the interior of the Earth toward its surface.

This book is the result of contributions from several experts and researchers worldwide. Due to its important utilization and future prospects, various interesting topics of research related to geothermal energy explorations are covered in this book. The topics cover the following important aspects: principles of geothermal power generation using LEGE-ORC technology; geothermal exploration on the slate formation of Taiwan; geothermal potential of the global oil industry; development stages of a geothermal project; and space cooling by ground source heat pump in Tropical Asia. It is hoped that the book will become a useful source of information and basis for extended research for researchers, academics, policy makers, and practitioners in the area of renewable geothermal energy explorations.

I would like to thank all chapter authors for their efforts and the quality of the chapters presented. Also, I would like to thank Ms. Dolores Kuzelj from IntechOpen for her outstanding efforts and patience in managing the publication process of this book.

This book contains five chapters. An introductory chapter (Chapter 1) highlights the principles of geothermal power generation using LEGE-ORC technology and presents a summary of the following book chapters.

Chapter two introduces the exploration results of a geothermal reservoir located in the slate formation of Taiwan using geological, geophysical, and geochemical methods. The hot springs, gas fumaroles with sulfur precipitation, hydrothermal alteration, and high surface temperature on rock bodies are the thermal manifestations in the Chingshui geothermal field. Geophysical surveys including geomagnetic, gravity, resistivity measurements (Transient Electro Magnetic (TEM) and Magnetotelluric (MT)) and micro-seismicity have been applied to detect the subsurface geological structures. A model with two geothermal reservoirs has been proposed underneath the Chingshui geothermal field. One reservoir is shallower at a depth of less than 3,000 m, while the other is deeper at depths ranging from 4,000 m to 8,000 m. Moreover, abundant micro-seismicity is distributed at the top of the deep reservoir to infer a high-temperature hydrothermal system with frequent hydraulic fracturing that induces micro-seismicity.

Chapter three covers useful and practical aspects on the geothermal potential of the global oil industry. Today, dedicated geothermal plants produce about 13 MW globally. The oil industry could generate much the same amount of power by processing the co-produced water from aging fields to produce power using Organic Rankine Cycle engines. If co-produced water was considered as an asset rather than a waste product, it is likely that oil fields would be operated longer with ever-increasing water oil ratios and the geothermal utility of the fields would increase. Utilization of the geothermal resources from oilfields would also help curb emissions by reducing the amount of co-produced gas used for power generation and ultimately allow more oil to be recovered from the fields because they could be run economically for longer with even lower oil production rates.

Chapter four primarily discusses the development stages of a geothermal project. A geothermal project is constituted by two big stages: the exploration and the exploitation. Each of these stages has sub-stages whose results allow for defining of the feasibility of a geothermal project to reach its stage of construction and operation of a power generation plant. The first stage is the recognition of the area, its limitation to the target, and the elimination of external factors until the geothermal zone is defined with characteristics to be commercially exploited. The main studies and analysis that have to be applied during the exploration stage are listed, until conclusion with the pre-feasibility report, which is the major indicator to continue with the project or suspend. The major risks in the exploration stage are presented and these are related to the studies that are carried out on the surface, inversely at this stage the costs can be considered low. The main product of the entire exploration is the selection of sites to drill three or four initial wells. The well provides a direct view of the reservoir; its depth, its production thicknesses, its thermodynamic parameters, and its production characteristics. The production parameters must be characterized in order to establish the well initial conditions, as well as its exploitation designs during the operation stage. Throughout the operation stage of each well, the monitoring of production histories, their characterization, their declination trends as well as the estimation of its useful life and remaining reserve should be continued. During the start-up phase of the plant, only operating and maintenance costs are considered, because the operating fuel of the geothermal plant is a natural and renewable resource. A review of the direct uses for the integral use of the resource is made during the exploitation stage as well as in low enthalpy systems.

Finally, in chapter five, space cooling by a ground source heat pump (GSHP) in tropical Asia is presented and discussed. The possibility of GSHP application in tropical Asia is studied based on groundwater temperature survey data in the Chao-Phraya plain, Thailand and in the Red-river plain in Vietnam compared with atmospheric temperature data. As a result, in most cities in these areas, the subsurface temperature is lower than atmospheric temperature in the daytime for most months. Thus, it is suggested that shallow underground layers may be used as a "cold" heat source at least in the daytime. Therefore, experimental operations of GSHP have been conducted in Thailand, Indonesia, and Vietnam to confirm its applicability. For better designing of the GSHP system in monsoonal Asia, numerical modeling of the regional groundwater flow and heat exchange simulation at the target location is recommended to determine the proper length and depth of the borehole heat exchanger. In the case of a semi-tropical region, heating by GSHP may create a new market for comfortable living standards. The studies shown here may be applied to other tropical and semi-tropical regions in the world.

Dr. Basel I. Ismail, P.Eng.
Chair and Associate Professor,
Department of Mechanical Engineering,
Faculty of Engineering,
Lakehead University,
Thunder Bay, Ontario, Canada

Chapter 1

Introductory Chapter: Power Generation Using Geothermal Low-Enthalpy Resources and ORC Technology

Basel I. Ismail

1. Introduction

The natural heat energy produced from the Earth is called "geothermal heat energy." It is proved to be an environmentally friendly clean energy source that could significantly help to mitigate GHG emissions when used for power generation [1]. The source of geothermal energy is the continuous heat energy flux flowing from the interior of the Earth toward its ground surface. The geothermal resources of the Earth are vast. For instance, the amount of geothermal energy stored at a depth of approximately 3 km is estimated to be 1,194,444,444 TWh which is much larger compared to all fossil fuel resources combined, whose energy equivalent is estimated to be 1,010,361 TWh [1, 2]. Unlike other renewable energy resources, geothermal energy has distinct characteristics; namely, it is available and stable at all times throughout the year, has an inherent storage capability, and is independent of weather conditions [1]. It was estimated that the world net electricity demand is going to increase by approximately 85% from 2004 to 2030, increasing from 16,424 TWh (in 2004) to 30,364 TWh (in 2030), so that the utilization of geothermal energy for electric power production continues to be a promising solution, more particularly with the new discoveries of innovative technological methods of power generation cycles and drilling technologies. The utilization of geothermal energy resources can also be used for direct heating applications [1]. In general, the utilization of geothermal energy is divided into two parts: (a) electricity generation and (b) direct heating (non-electrical) applications. Recently, this form of renewable energy source has grown in 25 countries, with installed geothermal-electric capacity up to 11 GW_e in 2010 [1, 2], and is increasingly contributing to the electric power supply worldwide [1–3].

Geothermal energy resources vary from one geographic location to another, depending on the depth, temperature, and pressure of the geothermal resource, the abundance of ground water, and the underground chemical composition. Geothermal energy resources typically differ in temperature from about 50 to 350°C. The high-temperature geothermal resources (with temperature >200°C) are typically found in volcanic regions and island chains, whereas the medium-temperature (150–200°C) and low-temperature geothermal resources (<150°C) are usually found widely in most continental regions and by far the most commonly available geothermal resource [2, 3]. The increase in temperature with depth in the Earth's crust can be expressed in terms of what is known as the geothermal temperature gradient. Down to the depths accessible by drilling with state-of-the art technology

IntechOpen

(over 10 km), the average geothermal gradient is about 2.5–3.0°C/100 m [2]. For example, at depth around 3 km below ground surface, the estimated temperature is approximately 90°C. There are, however, regions in which the geothermal temperature gradient is far from the average value. For instance, in some geothermal areas, the gradient is 10 times the average value due to composition of these areas and geothermal structure [2]. It was also reported that the emissions of GHG from geothermal power plants, in general, constitute less than 2% of the emission of these gases by fossil-fuelled power plants [2]. To comply with future energy demands, potential renewable energy sources should meet the following criteria: (1) the sources should be technically and economically accessible, (2) the sources should be large enough to sustain a long-lasting energy supply to generate the required electricity for the country, (3) the sources should be environmentally friendly and thus should be low GHG emitters in order to make significant contribution to global warming mitigation, and (4) the sources should have a wide geographic distribution [1–3]. Low-enthalpy geothermal energy (LEGE) resources extraordinarily satisfy all of these important criteria. This vast LEGE resource has already been utilized for electric power generation by some countries, such as the USA, Mexico, the Philippines, Indonesia, Austria, Germany, and Iceland [2]. The installations of several commercial LEGE electrical power systems in these countries have substantially proved the ability of low-temperature geothermal fluids to generate electricity [2]. In most developing countries, LEGE resources have not received much attention for electricity production. The key reason for not utilizing these resources by most developing countries and several industrialized countries for commercial exploitations is that they are not considered as cost-effective for generating electricity [2]. Recent increases in the cost and uncertainty of future conventional energy supplies for power generation are improving the attractiveness of LEGE resources. Generating electricity from medium- and low-enthalpy geothermal resources (water-dominated resources) can be effectively achieved using a binary cycle method which is also known as organic Rankine cycle (ORC) method [1, 2]. LEGE-ORC technology has virtually no GHG emissions to the atmosphere [1–3] and is a promising technology due to its simplicity and its limited number of components, all of them being very common and commercially available.

In this introductory chapter, the basic concept of LEGE-ORC binary power technology using LEGE heat sources is introduced, and its potential applications and limitations for small-scale geothermal power generation and its relevant environmental and economic considerations are presented and discussed.

2. Basic concept of LEGE-binary ORC technology for electrical power generation

A schematic diagram showing a LEGE-ORC binary-fluid system used for electric power generation is shown in **Figure 1**.

In this system, the first (primary) fluid being the geo-fluid (brine) is extracted from the LEGE resource through the production well. The geo-fluid carries the heat from the liquid-dominated resource and efficiently transfers this heat to the low boiling point (BP) organic working fluid (the secondary/binary fluid) using an effective heat exchanger. Shell-and-tube heat exchangers are widely used in these applications [1, 2]. The ORC is a thermodynamic Rankine cycle that uses the organic working fluid instead of steam (i.e., water). In this binary-fluid system, the low boiling point organic liquid absorbs the heat which is transferred by the geothermal fluid and boils at a relatively much lower temperature (compared to water) and as a result develops significant vapor pressure sufficient to drive the axial flow

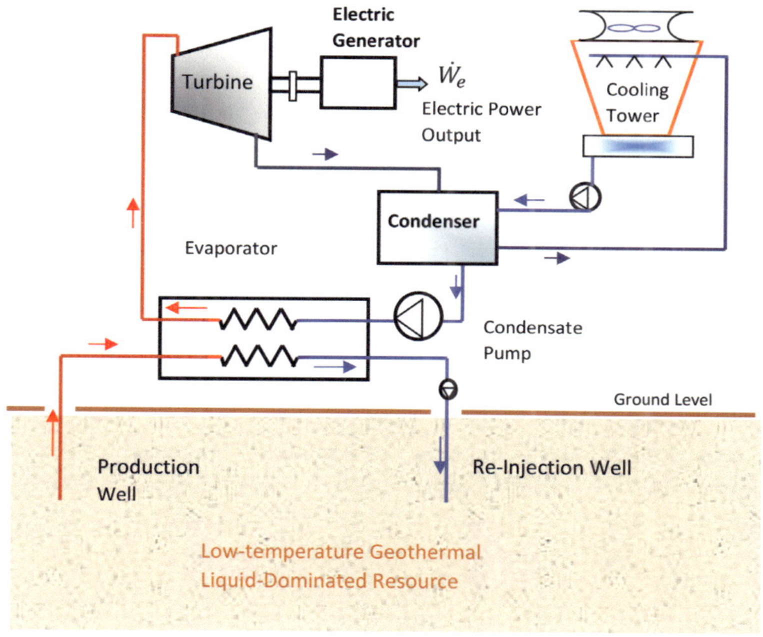

Figure 1.
A schematic showing the basic concept of LEGE-ORC technology for power generation [1].

or radial inflow turbine. The turbine is coupled to an electric generator which converts the turbine mechanical shaft power into electrical power. The organic working fluid expands across the turbine and then is cooled and condensed in the condenser before it is pumped back as a liquid to the heat exchanger using a condensate pump to be re-evaporated, and the power cycle repeats itself. One of the most important performance criteria in LEGE-ORC power generation technology requires the optimal selection of the ORC organic working fluid. Organic fluids used in binary ORC technology have inherent feature (compared to water) and that is they have low boiling temperature and high vapor pressure at relatively low temperatures, compared with steam (water) [1]. It was noted in [1] that typical ORC organic fluids may include pure hydrocarbons (e.g., pentane, butane, propane, etc.), refrigerants (e.g., R134a, R218, R123, R113, R125, etc.), or organic mixtures. The optimal energy conversion performance of a LEGE-ORC power generation system depends mainly on the type of organic fluid being used in the system. The selection of the type of organic fluid is typically based on the following criteria [1, 2]:

- The ORC organic fluid should result in high thermal efficiency by allowing maximum utilization of the available low-temperature geothermal heat source.
- It should be safe (non-flammable and nontoxic) and non-corrosive.
- It should not react or disassociate at the pressures and temperatures at which it is used.
- It should have suitable thermal stability and high thermal conductivity.
- It should have appropriate low critical temperature and pressure.
- It should result in low maintenance.

- It should have a low boiling temperature and should evaporate at atmospheric pressure.
- It should be environmentally friendly and less in ozone depletion potential (ODP) and global warming potential (GWP).
- It should have small specific volume and low viscosity and surface tension.
- It should lead to optimum design and cost-effectiveness of the ORC system.

It was also mentioned in [1] that many binary ORC fluids may not meet all these criteria but the selection of the organic fluid should be optimized, in terms of the above requirements, while meeting the demanded power generation. In general, binary ORC systems exhibit great flexibility, low maintenance, and high safety (installations are perfectly tight). It was also reported in [1] that the selection of suitable organic fluids for application in binary ORC systems for generating electricity still deserves extensive thermodynamic, technical, and feasibility studies.

3. Energy conversion efficiency of LEGE-ORC technology

In Ref. [1], the theoretical overall performance of LEGE-ORC binary systems can be evaluated using the fundamental thermal efficiency of a heat engine, given as:

$$\eta_{th} \equiv \frac{\dot{W}_{net,out}}{\dot{Q}_{geo,in}} \tag{1}$$

and also

$$\eta_{th} \equiv 1 - \frac{\dot{Q}_{cond}}{\dot{Q}_{geo,in}} \tag{2}$$

where $\dot{W}_{net,out}$ is the net power output delivered by the geothermal power system (in kW), $\dot{Q}_{geo,in}$ is the thermal power supplied by the geo-fluid from the available geothermal resource (in kW), and $\dot{Q}_{cond}$ is the thermal energy rejected in the condenser (in kW). For quick estimate purposes, a correlation is proposed [1, 2] to calculate the actual net power output (with rough accuracy), given by

$$\dot{W}_{net,act} = \left(\frac{1}{278}\right)[(0.18T_{geo,in} - 10)\ \dot{Q}_{geo,in}] \tag{3}$$

Substituting Eq. (3) in Eq. (1), the estimated thermal efficiency of the low temperature-based geothermal power generation system, as a function of geo-fluid inlet temperature (in °C) at the production well, is given by

$$\eta_{th} \approx \left(\frac{1}{278}\right)(0.18T_{geo,in} - 10) \tag{4}$$

The thermal efficiency as a function of the low-temperature geothermal heat resource temperature, T_H (in K), and ambient temperature, T_a (in K), is given by

$$\eta_{th} \cong \left(\frac{58}{100}\right)\left(\frac{T_H - T_a}{T_H + T_a}\right) \tag{5}$$

The estimated net power output delivered by the geothermal power system can also be determined using [2]

$$\dot{W}_{net,out} \cong 2.47\dot{m}_{geo}\left(\frac{T_H - T_o}{T_H + T_o}\right)(T_H - T_C) \tag{6}$$

4. Installations and land usage aspects for using LEGE-ORC technology

LEGE-ORC power generation systems are typically constructed and installed in small modular, compact (and in some cases mobile) units of a few hundred kW_e to a few MW_e capacities [2]. These units can then be integrated to form power plants of 10–50 MW_e which are considered to be small power plants. It was estimated that 1 MW_e power plant could serve about 20,000 households assuming that the demand for electricity per person at off-grid sites will be of the order of 500 W [1, 2]. The convenience of the small power plants is most evident for areas without ready access to conventional fuels, but with access to LEGE resources, and for communities that it would be very expensive to connect to the national electric grid [2–4]. Basic LEGE-ORC power generation systems typically include the following list of equipment [2]:

- Downwell pumps
- Geo-fluid (brine) supply system (sand removal system)
- Heat exchangers (evaporator/condenser, shell-and-tube type)
- Turbine generator and controls
- Condensate pump
- Piping system
- Heat rejection system (cooling tower)
- Backup electric power system
- Fire protection system (if the ORC working fluid is flammable)

The area required to support a geothermal power facility, including the well field, access roads, and auxiliary buildings, depends on the geothermal power plant power rating, the properties of the geothermal reservoir fluid, and the piping system selected for collecting the geo-fluid from the production wells and disposing the waste brine to the reinjection wells. The geothermal power facility should be installed close to the production wells to avoid thermal and hydraulic losses caused by long geo-fluid pipelines. The pipelines used to transport the geo-fluids are usually mounted on stanchions, run along service roads, and incorporate vertical and horizontal expansion loops [2]. It should be noted that a low-enthalpy geothermal power plant typically requires (per MW_e) 5% of the area needed for a solar thermal power plant and 2% for a solar PV power plant located in the best solar insolation area in the USA. It was reported also in [2] that a 20 MW_e geothermal binary power plant (excluding wells) requires a land area of 1415 m^2/MW_e.

5. Environmental aspects of using LEGE-ORC technology

Geothermal energy is relatively free from pollution problems and considered to be a clean technology. It seems to have the largest technological potential compared to other renewable energy sources [1–4]. Emissions are typically zero when LEGE reservoirs are utilized using ORC binary technology, since all of the produced geo-fluid is injected back into the reservoir. Some chemical constituents of geothermal water (e.g., trace metals) may need to be monitored in case their concentrations exceed permitted pollution limits. One of the effective ways of getting rid of hazardous chemicals is reinjection. Low-enthalpy geothermal binary power generation systems are far less environmentally intrusive than alternative power generation systems in several respects, for example, they are essentially zero-GHG emission systems and have low land usage per installed megawatt [2]. As far as physical environmental effects, geothermal projects may cause some kind of disruption activities as other same size and complexity of civil engineering projects. Also, the locations of excavations and sitting of boreholes and roads will have to be taken into account; soil and vegetation erosion, which may cause changes in ecosystems, has to be watched. There is considerable noise involved during geothermal drilling, construction, and production phases of development, so that protection has to be ensured for residents, and some permanent noise-reduction measures may need to be considered. It should be noted that many geothermal developments are in remote areas where the natural level of noise is low and any additional noise is very noticeable [1, 2]. At the social-economic level, the construction of geothermal power installations involves a temporary increase in employment, and the building of new roads may open up areas and possibly increase tourism, especially if natural geothermal manifestations are left intact. Geothermal power facilities utilize low-source temperature to produce the primary thermal energy for conversion to power production, and therefore the waste heat per MW_e of electricity generated is much larger than others types of power generation [2]. Appropriate measures should be applied to prevent leakage of the binary working fluid from ORC power generation units to the environment [1, 2]; normally the installations of these units are made perfectly tight to meet high safety requirements.

6. Economic aspects of using LEGE-ORC technology

Generating electricity using geothermal ORC technology is very cost-effective and reliable [1, 2]. **Table 1** [2] compares the unit cost of electricity generated from low enthalpy-based small power plants. Moreover, the unit cost of electricity from small-scale geothermal plants (<5 MW_e) is much lower than the average cost of

Net power (kW_e)	Capital cost (US$/net kW_e)			O and M cost (US$/year)
	Resource temperature (°C)			
	100	120	140	
100	2786	2429	2215	21,010
200	2572	2242	2044	27,115
500	2357	2055	1874	33,446
1000	2143	1868	1704	48,400

Table 1.
Unit cost of electricity generated from LEGE-based small power plants [2].

DOI: http://dx.doi.org/10.5772/intechopen.84390

0.25 US$/kWh supplied through diesel generators [2]. The total investment for a geothermal power plant mainly includes the following: (1) cost of drilling, (2) cost of exploitation, (3) operating and maintenance costs, and (4) cost of power plant (capital cost of design and construction) [1, 2]. The first two types are referred to as subsurface costs, whereas the other two are referred to as surface costs. For small-scale geothermal power plants (<5 MW_e) utilizing LEGE resources, the subsurface cost typically accounts for approximately 30% of the total investment costs, whereas the surface cost accounts for the remaining 70%. For larger geothermal power plants (>5 MW_e) utilizing high-enthalpy resources, the surface cost is a small part of the total cost of the project. The major cost involved in larger plants seems to be the subsurface cost. Generating electricity using LEGE-ORC technology is very reliable due to its advanced technological aspects [2]. However, the maintenance costs and shutdowns could be reduced when the technical complexity of the plant is on a level that is accessible to local technical personnel or to experts who are readily available [2]. Also, reducing operations and maintenance costs can be achieved through advanced methods to control the geo-fluid (brine) chemistry and the use of more robust instruments for tighter controls [1, 2]. Geothermal ORC power generation plants are normally constructed and installed in small modular power generation units. These units can then be linked up to create power plants with larger power production rates. Their cost depends on a number of factors, but mainly on the temperature of the geothermal fluid produced, which influences the size of the ORC turbine, heat exchangers, and cooling system. The total size of the plant has little effect on the specific cost, as a series of standard modular units is linked together to obtain larger power capacities [2]. The modular units have a satisfying economic efficiency, because modular construction reduces installation time and costs. Ultimately, the economic viability of the geothermal power plant depends on its ability to generate revenue in the long term. This is because of relatively lower unit cost of electricity generated by geothermal plants and the future power-sale factor. This means that geothermal power plants, in general, are normally built and designed to serve for more years in order to get to the point where they could pay back for the investment cost and start to generate revenue and ultimately become highly cost-effective [2].

Author details

Basel I. Ismail
Department of Mechanical Engineering, Lakehead University, Canada

*Address all correspondence to: baseliai@gmail.com

References

[1] Ismail BI. Chapter 18: Power generation using nonconventional renewable geothermal & alternative clean energy technologies. In: Carayannis EG, editor. Planet Earth 2011—Global Warming Challenges and Opportunities for Policy and Practice. Rijeka, Croatia: InTech Open Access Publishing Company; 2011. ISBN: 978-953-307-733-8

[2] Ismail BI. Advanced electrical power generation technology using renewable & clean low-enthalpy geothermal energy sources. Recent Patents on Mechanical Engineering. 2011;**4**(2):168-179

[3] Ismail BI, editor. Advances in Geothermal Energy. Rijeka, Croatia: InTech Open Access Publishing Company; 2016. DOI: 10.5772/60623. 174 p. ISBN: 978-953-51-2241-8

[4] Ismail BI. Chapter 13: ORC-based geothermal power generation and CO_2-based EGS for combined green power generation and CO_2 sequestration. In: Arman H, Yuksel I, editors. New Developments in Renewable Energy. Rijeka, Croatia: InTech Open Access Publishing Company; 2013. ISBN: 978-953-51-1040-8

Chapter 2

Geothermal Explorations on the Slate Formation of Taiwan

Sheng-Rong Song and Yi-Chia Lu

Abstract

Currently, over 90% operated geothermal power plants are distributed in the volcanic- or magmatic intrusion-related geological systems. Only a few cases are done in metamorphic terranes, especially on the slate formation. Taiwan is located at the ring of fire and is famous for the young orogenic belt, which has wide distributions of rapid uplifting terranes with few active volcanoes. The metamorphic rocks, for example, schist and slate formations with high geothermal gradients, are occurring in the major mountain range. This chapter introduces the techniques or methods we used for geothermal exploration in the slate formation of the Chingshui geothermal field of Taiwan, where a 3-MW pilot geothermal power plant had been installed in 1983 and operated for 12 years.

Keywords: Chingshui geothermal field, Taiwan, slate formatiom, calcite veins, clumped isotope

1. Introduction

Energy is the essential lifeblood of today's national economy. Diversification of energy sources is reasonable and important in establishing policy for all countries in the world. Among various energy sources, geothermal energy offers a naturally free resource and less dependence on fossil fuels. Taiwan is a country in which the energy is very scant, and almost over 99% consumed ones, most of the fossil fuels, are imports from outside of this country [1]. Taipower, the only commercialized power company in Taiwan, has installed capacity of 40.79 GWe, which produces the gross electricity of about 219.2 billion kWh [2]. They are produced by fossil fuels including oils, coals, and natural gases, nuclear power, hydropower and renewable energy, etc. Among them, the majority is the fossil fuels (72.78%) followed by nuclear power (18.61%), while the renewable energy only occupies 2.86%. There are four nuclear power plants in Taiwan, namely, No. 1, No. 2, No. 3, and No. 4, three of which are in operation and the fourth is under construction, but the latter is sealed now due to it being considered as non-safe after the 2011 Fukushima nuclear disaster in Japan and because of the nuclear-free policy. Moreover, those three operating ones are old, more than 30 years old, and will be decommissioned from 2018 to 2025. The electricity will be less than 18.61%, about 40.79 billion kWh per year, and there is a need to find alternatives for them in the next 10 years in Taiwan. It, thus, provides an opportunity for developing geothermal energy.

Taiwan is located at the ring of fire and is famous for the young orogenic belt (**Figure 1**), which has wide distributions of rapid uplifting terranes with few active volcanoes [3, 4]. Three geothermal play types have been distinguished in

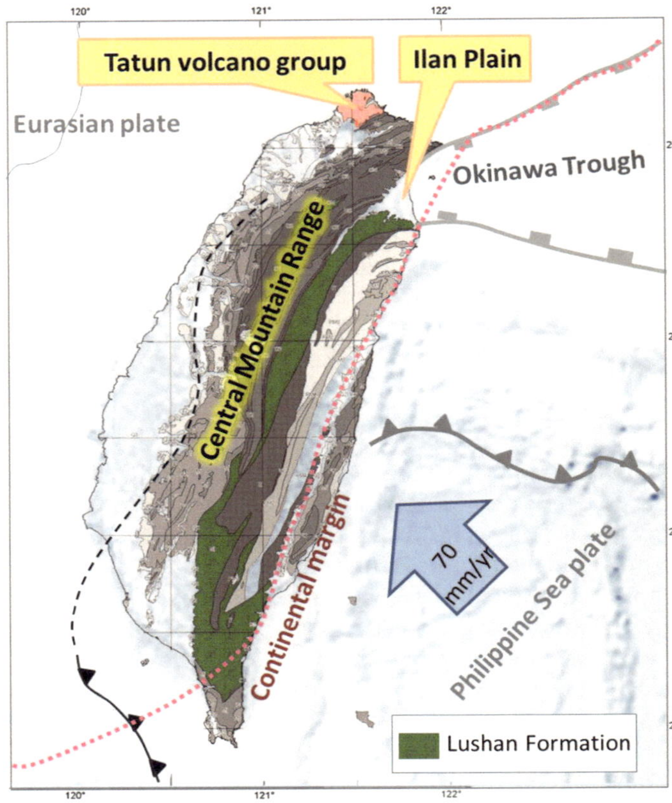

Figure 1.
The tectonic framework of Taiwan.

terms of geological controls [5, 6]. They are the magmatic-volcanic field type, the extensional domain type, and the orogenic belt/foreland basin type, which are correlated to the Tatun volcano group, the Ilan Plain, and the Central Mountain Range, respectively (**Figure 1**). Lot of geothermal explorations have been done in this island since the 1960s, and two pilot geothermal power plants had been built up in slate formation in northeast Taiwan [7]. In 2005, the Bureau of Energy of Taiwan revisited the slate formation for geothermal exploration and developed a plan for future power generation in the Chingshui geothermal field. The Ministry of Science and Technology (MOST) (formerly known as the National Science Council, NSC) of Taiwan initiated and promoted the geothermal explorations and developments as a major national energy project (NEP I and II) since 2008. The works include more precise geological, geochemical, and geophysical surveys with drillings [8]. Currently, over 90% of the on-working geothermal power plants are operated in the volcanic- or magmatic intrusion-related systems [9]. There are only few cases of metamorphic terrane, especially on the slate formation. The aims, thus, of this article are to introduce the techniques or methods we used for geothermal exploration in the slate formation of the Chingshui geothermal field of Taiwan.

2. Geological background

Taiwan belongs to the ring of fire and is famous for active orogeny (**Figure 1**). Presently, the Philippine Sea plate is moving toward WNW at about 70 mm/yr. [10], and it is believed that the mountain-building process is still ongoing [11, 12]. A dominant collision zone frequently inducing folding and fault thrusting in the area may exist in central Taiwan and cause rapid uplifting rate being 6 mm per year [13]. Tectonically, the Philippine Sea plate is riding up over the continental shelf of the South China Sea in southern Taiwan. Moreover, an oceanic part of the Eurasian plate is subducting beneath the Philippine Sea plate along the Manila trench, which results in the bulldozing of shelf sediments both upward and westward. This island of Taiwan has been created in the last 5 million years [3, 14, 15], and rapid crystal movements and widely distributed active structures make up the geological characteristics of this young tectonic entity [16–19]. In the north, the Philippine Sea plate subducts underneath the Eurasian plate, leading to the formation of Ryukyu arc (**Figure 1**). The Okinawa Trough is a back-arc basin, which extends from southwest Kyushu Island (Japan) to the Ilan Plain, which is the southwest-most tip of it. Three stages of opening have been identified since 15 Ma, and the latest phase of extension occurred in the southwestern part of the Okinawa Trough, which is characterized by normal faults with vertical offsets since the late Pleistocene [20–22]. The age of the normal stress affecting the Ilan Plain may be in the latest Pleistocene; based on the thermoluminescence (TL), age of 7.0 ± 0.7 ka obtained from a siltstone xenolith was found at Kueishantao, an offshore volcanic island 10 km away from the coast of Ilan Plain [23]. Several active volcanoes have been identified in inland and offshore of this island, and high uplift mountain range occurs in the eastern and central parts of Taiwan. Those tectonic settings provide a very good environment rich in geothermal energy.

The slate formation, named the Lushan Formation, is widely distributed in the Backbone Range belt of Central Range, Taiwan (**Figure 1**). It is composed of argillite and slate of early to middle Miocene. The type locality of this formation is located in the Lushan area, east of Wushe in Nantou County. Miocene foraminifers were found in slates and marly nodules in this formation, which extend for an east–west width of at least 14 km on the western slope of Central Range [24, 25]. The Lushan Formation consists largely of black to dark-gray argillite, slate, and phyllite with occasional interlayer of gray sandstones. The estimated thickness is several thousand meters. This formation is exposed in north Taiwan at the mouth of the Lanyang River in the Ilan County and extends southward along the crest zone of the Central Range through Hohuashan, Nengkaoshan, to Hsiukuluanshan for a length of approximately 150 km and a width of several to 10 or more kilometers [26]. South of the Yushan Mountain, the Lushan Formation occurred east of the Laonong River southward to the eastern mountains of the Pingtung valley down to the Hengchun Peninsula. It also crops out in southeastern Taiwan in the area of Chihpen and Tawu [26].

3. Geology of Chingshui geothermal field

The Chingshui geothermal field is located in the valley of Chingshui stream, southwest of Ilan Plain, northeast Taiwan (**Figure 2**). The rock hosting the geothermal field is the Miocene Lushan Formation, consisting dominantly of argillite/slate with intercalated thin meta-sandstones [27, 28]. Several hot springs with minor hydrothermal alterations are the important thermal manifestations, which they crop out along the riverbed and rock cliffs in this area (**Figure 3A**). Temperatures and TDS of them range from 34°C to over 95°C and from 896 to 1500 ppm, respectively. Hydrothermal minerals include calcite, aragonite, dolomite, strontianite, amorphous silica, burbankite, kaolinite, sulfur, jarosite, tschermigite, and gypsum.

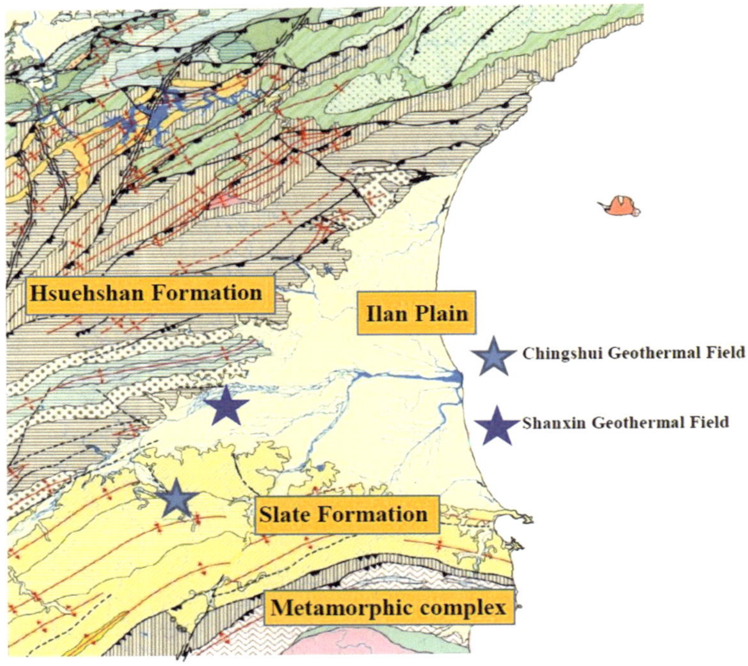

Figure 2.
The Chingshui geothermal field is located in the valley of Chingshui River, which is in the southwestern Ilan Plain.

These field investigations and subsurface geological reports indicate that the Lushan Formation in this area is composed predominantly of dark-gray and black slates with thinly layered meta-sandstones, which can be divided into three members. They are the Jentse member, the Chingshuihu member, and the Kulu member. The upper Jentse member consisting of mainly light meta-sandstone intercalated in dark-gray slates, while the lower Chingshuihu and Kulu members consist mostly of slates [27, 28]. Two regional folds according to the vergence from minor folds in the thin-bedded sandstone have been identified in the Chingshui geothermal field [29]. The synclinorium axis is located within the Jentse member across the upstream region of the Chingshuichi River. The anticlinorium axis is located across the downstream region of Chingshuichi in the Chingshuihu member. Both fold axes were offset by the Chingshuichi Fault.

Several faults, including the Xiaonanao and Chingshuichi faults and a few small unnamed ones, cut through rock bodies in Chingshui area (**Figure 3A**). The Xiaonanao Fault is a thrust fault with wide damage zones that are rich in quartz veins on the outcrops along the Chilukeng River (**Figure 4**). The fault stretches to the east where an upside-down sequence of strata appears at the Shimen River [28]. The Chungshuishi Fault is a south–north strike-slip one inferred from geophysical data [30, 31]. The geothermal field, therefore, cropped out follows the surface trace of the suspected Chingshuichi Fault along the Chingshuichi Valley [27, 30, 32] (**Figure 3A**). Those normal faults, therefore, in the area were formed during the Pengli orogeny in the late Pliocene [33]. However, some outcrops of other faults cannot be found on the surface after heavy weathering and erosion in the area.

There are many parallel normal faults with strike-slip component gouges (inferred by slickenside direction) approximately 200 m long along the confluence of the Chingshui River and the Chilukeng River [34] (the star mark at **Figures 3A** and **5A**).

(A)

(B)

(C)

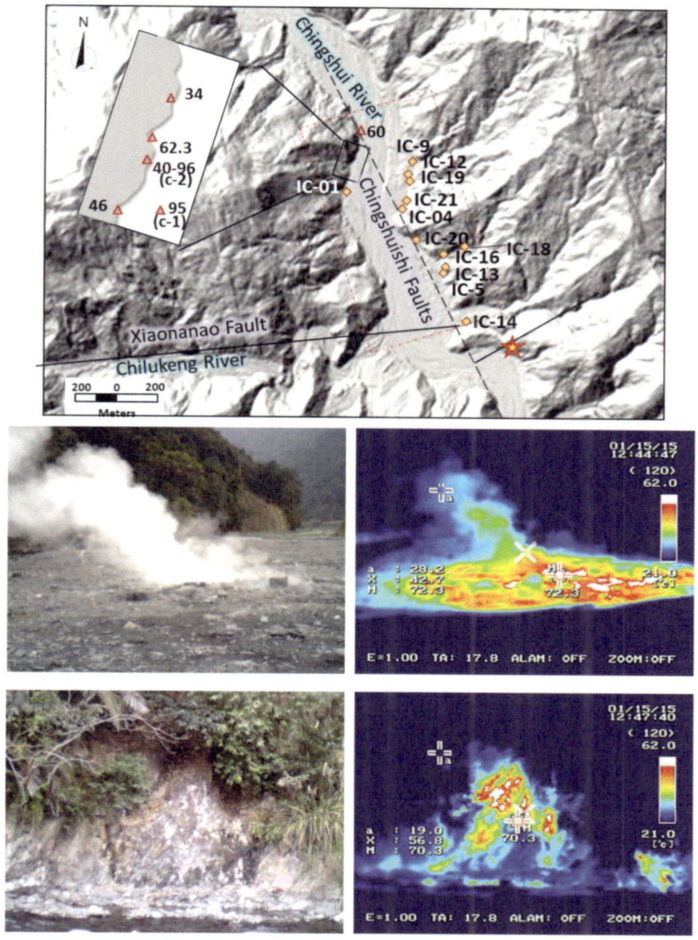

Figure 3.
(A) Distribution of surface temperature on rocks along the Chingshui River. (B) Thermal (left) and optical (right) images of hot springs along the riverbed. (C) Thermal (left) and optical (right) images of hot springs and fumaroles along the rock outcrops.

Figure 4.
The Xiaonanao Fault is a thrust fault with wide damage zones that are rich in quartz veins with euhedral pyramidal crystals on the outcrops along the Chilukeng River.

The fault gouge strikes ranging from N55°E to N75°E and dips from 30°N to 80°N. The mineral assemblage for these white veins is predominantly composed of calcite with minor quartz (**Figure 5B**).

Figure 5.
(A) A normal fault (white line), located at the confluence of Chingshui River and Chilukeng River (star mark at **Figure 3A***), is approximately 2 m wide with a steep dip and contains blocks of quartz veins. (B) The vein in the damage zone, cut by steep normal fault, is predominantly composed of calcite with minor quartz.*

4. Geophysical explorations

Several geophysical exploration methods have been applied to the Chingshui geothermal field. They were geomagnetic, gravity, and resistivity surveys including transient electromagnetic (TEM), magnetotelluric (MT), and microseismicity.

4.1 Geomagnetic survey

Magnetic surveys record the spatial variation in the Earth's magnetic field, which the magnetic properties of rocks are measured. Generally, magnetic susceptibility is summarized as the various amounts of minor accessory minerals in all rocks, which contain iron-rich phases such as magnetite, pyrrhotite, and hematite. Magnetite is the most important magnetic mineral, because it is not only very common but also has relatively high magnetic susceptibility. A geomagnetic survey measures the changes in the amounts of magnetic minerals as well as associated rock types. A magnetic map, thus, helps locate mineral deposits by identifying specific rock types and geological features [35].

A total of 425 stations for geomagnetic survey in the Ilan Plain were performed in 1978 by Yu and Tsai [31]. They found an obvious WE high magnetic anomaly which is located between Ilan and Luodong, and they interpreted that this anomaly could be the magma intrusion as dikes underneath the Ilan area. Meanwhile, a low-magnetic zone between Chingshui and Hanhsi was observed by Tong et al. [30], which reprocessed the old data and elucidated that it might be associated with the destruction of magnetite in the host rocks by hydrothermal alteration. The fluid for hydrothermal alteration in the Chingshui area was probably related to the magmatic source also supported by oxygen and carbon isotopes [34].

4.2 Gravity survey

The purpose of gravity survey is to detect mass materials underneath the places, which are not uniformly the same everywhere. They vary with the distributions of the dense materials below. A gravity survey is an indirect method to get the density property of subsurface materials. The higher the gravity values, the denser the rock, such as igneous body beneath.

A gravity survey with a total of 636 stations in the Chingshui geothermal field was completed by the ITRI in 1976 [36]. Although the variation in gravity is not significant due to the fact that the rock formation in survey area is predominantly composed of slates, the gravity on both sides of the Chingshui stream is apparently different [30], which may be related to fault cut through along the river. Euler deconvolution is a useful tool to interpret the gravity data rapidly and delineate structural contacts and depth estimation quickly [37, 38]. Based on the results of Euler deconvolution and known adjacent geological structures, several faults have been identified. They are the Xiaonanao, Chingshuihsi, and Kulu faults, which correspond with known faults introduced by Tseng [28] and Lin and Yang [33]. Meanwhile, three unknown faults, namely, the A-fault, B-fault, and C-fault, have also been recognized, which might be associated with the Niutou Fault with a SW–NE trend. The Chingshuihsi Fault was offset by the Xiaonanao Fault and the C-fault in the south and north of Chingshui, respectively, and the known geothermal field in this region is bounded by the C-fault and the Xiaonanao Fault [30, 33].

4.3 Resistivity surveys

Surface resistivity survey measures the electrical potentials in the ground around a current-carrying electrode depending on the electrical resistivity and distribution of the surrounding sediments and rocks. It applies an electrical current between two electrodes implanted in the ground to measure the difference of potential between two additional electrodes that do not carry current. The distribution of potential can be related theoretically to ground resistivity and their distributions for rock bodies, which are distributed in a horizontally stratified ground and the homogeneous masses separated by vertical planes [39]. Mineral grains in sediments and rocks are essentially nonconductive; except in some ores or metallic minerals, the resistivity of sediments and rocks is governed primarily by the amount of pore water, its resistivity, and the orientations of the pores. The differences of lithology have different resistivity, so the surveys can be used to detect bodies of anomalous materials or in estimating the depths of bedrocks [39]. Generally, since the resistivity of rocks is controlled primarily by the pore water conditions in a rock, the values cannot be directly interpreted in terms of lithology. However, zones of distinctive resistivity are associated with specific rock units on the basis of local field or drill hole information, and the surveys were used profitably to extend field investigations into areas with very limited or nonexistent data. Meanwhile, the resistivity surveys may be used as a reconnaissance method, to detect anomalies that can be further investigated by complementary geophysical methods and/or drill holes [39].

The ITRI conducted surface resistivity surveys, including the transient electromagnetic (TEM, the collinear and orthogonal dipole–dipole measurements) and magnetotelluric (MT) in the Chingshui area since the 1970s [30, 32, 40, 41]. They deployed both the transmitter and receiver dipole lengths at different distances, that is, 300–1300 m, separately for TEM. Four groups of apparent resistivity data sets were collected with four individual transmitter locations and were reprocessed and inverted with modern 2D and 3D inversion codes later [42, 43]. The inverted bipole-bipole results show the regional geological structures of the Chingshui area and reveal three vertical conductive structures, H, I, and J. They may correspond to the vertical Chingshuichi Fault, the stratigraphic boundary between the Chingshuihu member and Jentse member, and coincide to the Dachi Fault [28] and the Xiaonanauo Fault, respectively [43]. Hot springs crops out along the Chingshuichi Fault, suggesting that it might be the conduit to provide geothermal fluids.

The magnetotelluric (MT) survey is a method to use frequently and successfully for exploring geothermal reservoirs [44–46]. Many reports claim that geothermal fluid circulates along the fractures within the meta-sandstones or slates in the Chingshui geothermal field [27, 28]. To detect the geological structures

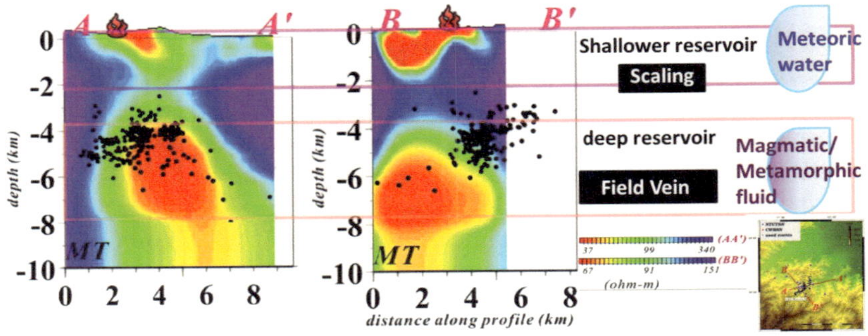

Figure 6.
Two-reservoir model with MT images and many microseismicity in the Chingshui geothermal field.

and reservoir underneath the Chingshui area, 33 broadband magnetotelluric data points were acquired by the ITRI in June 2006 [30]. Those data were processed and inversed for 1D and 2D images [30]. Moreover, the MT time series data were measured over 72 h at all stations to improve the data quality in 2014 and were processed using statistically robust algorithms from Jones et al. [47] for the MT 3D inversion [48].

The MT results could be achieved based on the understanding of the resistivity of various types of rocks in the survey area (**Figure 6**). A significant low-resistivity zone can be identified, which could be related to a clay-rich cap rock in the geothermal structure. The cap rock is about 1 km in width and is found at depths ranging between 200 and 1000 m [30]. The resistivity of the cap rock is about 14 Ωm [30], which is not as low as many other volcanic geothermal fields that have a low-resistivity cap layer of less than 5 Ωm. It is characterized by the Chingshui geothermal field having slate host rocks that are expected to be less reactive than the volcanic rocks. The cap rocks could be illite-rich slates, which got approved by recent drilling results [8].

The reservoir of Chingshui geothermal field is a typical fault-controlled fractures, which may be created by the several faults distributed in this area [30]. MT images show that the fractures associated with the geothermal reservoir are distributed from near surface to a depth of 1500 m toward the south in fault zones, which is similar to the identified production zone from the core drilling records. The size of the Chingshui geothermal reservoir is estimated to be about 9.54 × 107 m^3 and contains about 10 million cubic meters of geothermal fluids, based on the 3D model with a gross porosity of 0.1 and 100% saturation for the fracture zones [42]. Meanwhile, two geothermal reservoirs have been proposed underneath the Chingshui geothermal field according to MT images (**Figure 6**) [8, 34, 48]. One is shallower at a depth of less than 3000 m, while the other is deeper at depths ranging from 4000 to 8000 m. Moreover, abundant microseismicity occurred at the top of the deep reservoir [49]. This result leads us to infer that the deep reservoir may be a high-temperature hydrothermal system with frequent hydraulic fracturing occurring that induces microseismicity.

5. Geochemical explorations

Geothermal exploration provides abundant information for the location, nature, and origin of the geothermal waters in a geothermal system. Geochemical studies of geothermal fluids involve three main steps: (1) sample collection, (2) chemical analysis, and (3) data interpretation [50]. The results give the parameters that are sensitive to subsurface temperatures, salinity of the fluids, and gaseous chemical compositions in interesting areas. Meanwhile, the constituents of fluid chemistry are used to trace the origin and flow of geothermal waters, especially the stable isotopes, such as ^2H and ^{18}O, along with B and Cl being most important. Chemical constituents of rocks, for example, SiO_2, Na, K, Ca, Mg, and CO_2, are useful for estimating subsurface temperatures and potential production problems such as scaling and corrosion [50].

5.1 Geochemistry of fluids

Several hot springs cropped out on the surface, and 21 exploring and production wells have been drilled in the Chingshui geothermal field (**Figure 3A**). The temperatures and pH of springs and wells range from 48 to 99°C and 6.4 to 9.7 and from 180 to 230°C and 6.3 to 8.9, respectively [7]. The major gases of geothermal steam being about 20% in the Chingshui area are CO_2, H_2S, and others. Carbon dioxide is generally the major gas component often comprising more than 97% of all non-condensable gases, and its concentration increases with reservoir temperature. Hydrogen sulfide concentration is about 1% and commonly decreases as steam ascends to the surface due to reaction with wall rock, dissociation to sulfur, or oxidation to SO_4 [7].

The chemical and isotopic compositions of hot spring and wells are shown in **Figures 7** and **8**. Chemical compositions of cations and anions in the Chingshui geothermal fluids are pretty variable. Bicarbonate ($[HCO_3^-]$ = 500–3200 ppm) is the major anion in most geothermal waters, being with lower chloride ($[Cl^-]$ = 6.5–23.4 ppm) and sulfate ($[SO_4^{2-}]$ = 29–72 ppm), which are the bicarbonate fluid type based on Piper diagram (**Figure 7**) [51]. Sodium ($[Na^+]$ = 35–1235 ppm) is the major cation in most geothermal fluids, being with lower calcium and magnesium (both few ppm). Based on the Na-K-Ca geothermometer, the temperatures of thermal fluids in reservoirs range from 137 to 205°C. Silica (SiO_2) compositions of thermal fluids range from 83 to 413 ppm, which are correlated to the temperature from 127 to 214°C, respectively, by silica geothermometry [7]. However, the SiO_2 geothermometer is a more reasonable and suitable method than Na-K-Ca one for assessing reservoir fluid temperatures in the slate region in Taiwan in terms of experimental fluid–rock interactions on laboratory [52, 53].

Hydrogen and oxygen isotopic compositions for meteoric water along LangyongSi River (the mainstream of Chingshui River) were −70~ + 10‰ and −11 to 0‰, which may be due to rapid topographic change and strong monsoon effect in winter [54] (**Figure 8**). For the hot springs and thermal water, the δD and $\delta^{18}O$ values are −67~−32‰ and −9.2 to −4.4‰ and −57~−24‰ and −6.7 to −4.0‰, respectively. The wide ranges of isotopic compositions in hot springs may be partly attributed to wider geographic distributions of them and mixing of thermal water with meteoric water [54]. Plots of H^- and O^- isotopic compositions of thermal water on the local meteoric water line (MWL) show the close relationship with the meteoric water (**Figure 8**). Isotopic changes of geothermal water due to fluid–rock interaction were small with a maximum $\delta^{18}O$ shift of about 3‰ from the MWL. This small shift may reflect the slow fluid–rock interaction in terms of low permeability of the slate host rocks [54].

5.2 Isotopes of carbonate veins and scaling

The hot fluids in Chingshui geothermal field are characteristic of high concentration of HCO_3^-. When the geothermal reservoir is breached by tectonic activities or drilling for geothermal exploitation, CO_2 is oversaturated and can be released

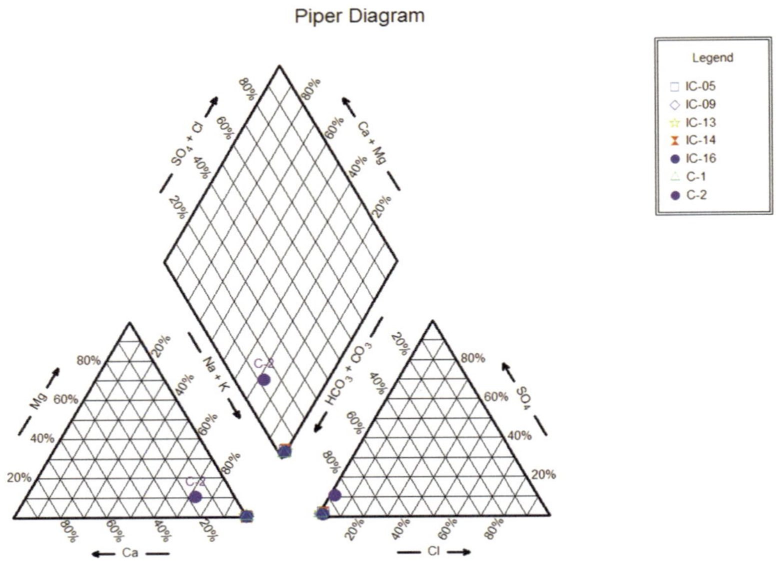

Figure 7.
The water compositions of Chingshui area are plotted on the Piper diagram showing the typical Na-bicarbonate fluid type.

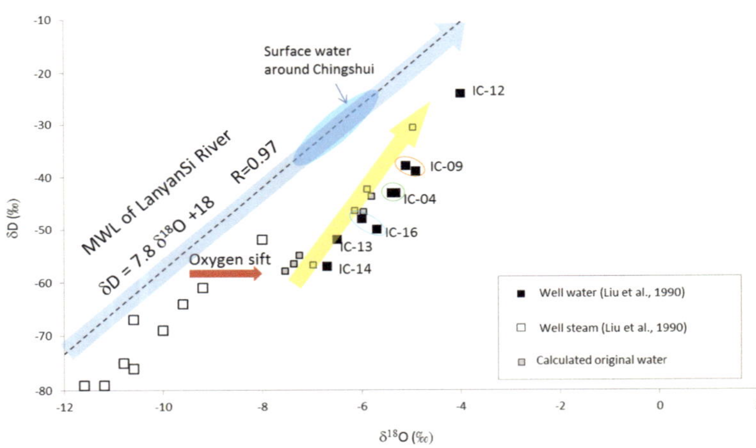

Figure 8.
Plots of H- and O- isotopic compositions of thermal water on the local meteoric water line (MWL) show the close relationship with the meteoric water.

quickly by depressurization causing the bicarbonate solution to oversaturate rapidly with pH increase and to precipitate carbonate minerals from thermal water immediately. The isotopic data from fracture-filling carbonate minerals have been found to be particularly useful to constrain the geochemical characteristics of fluid reservoirs and possible post-depositional and syntectonic fluid processes [55–59].

Two populations of ^{18}O values were recognized, −5.8 ± 0.8‰ VSMOW from scaling in the wells and −1.0 ± 1.6‰ to 10.0 ± 1.3‰ VSMOW from the calcite veins of outcrops (**Figure 9**), which are indicative of meteoric and magmatic fluid sources, respectively [34]. Meanwhile, two hydrothermal reservoirs at different depths have been identified by magnetotelluric (MT) imaging with microseismicity underneath this area [48, 49] (**Figure 6**). Two-reservoir model has been proposed: One is the shallow reservoir with fluids from meteoric water to provide the thermal water for scaling depositions inside the production wells, while the deep one supplies magmatic fluids mixing with deep marble decarbonization to precipitate the calcite veins near fault zones [34]. Helium isotope data from Cheng [60] also provided strong evidence of magmatic fluids from the deeper reservoir in the Chingshui geothermal field. The ratios of $^3He/^4He$ were 3.8–4.0 RA and 0.8 RA for the samples. These lines of evidence indicated the existence of a mantle-derived component in the Chingshui area, which may be derived from magmatic degassing; however, the lower helium isotope ratio of the other sample also implied a mixing between such a deeper, magmatic-related reservoir and a shallower, crustal-related one [34, 60].

The well IC-21 commenced drilling at May 2010. Upper 600 m was drilled into the hole and did not take any cores, just a cutting per 10 m. Whole coring raised 200 m in length between 600 and 800 m in depth and got over 95% core recovery. Lithologically, the 200 m cores are predominantly composed of dark-gray to black slates occasionally intercalated with argillites or meta-sandstones. There are many deformation structures, fractured systems, and veins in the cores (**Figure 10**). It is characteristic that many scaling minerals are irregularly filled up in the fractures, veins, and open cracks.

Three types of calcite crystal morphologies have been identified in the veins of the cores: bladed, rhombic, and massive crystals (**Figure 11**). Bladed calcites are generated via degassing under boiling conditions with a precipitation temperature of ~165°C and calculated $\delta^{18}O$ value of −6.8‰ to −10.2‰ VSMOW for the thermal water [61]. Rhombic calcites grow in low-concentration Ca^{2+} and CO_3^{2-} [62–64] meteoric fluids and precipitate at approximately ~180°C [61]. Finally, massive calcites coprecipitated with quartz in the mixing zone of meteoric water and magmatic or metamorphic fluids with calculated $\delta^{18}O$ value of up to 1.5 ± 0.7‰ VSMOW. Furthermore, the scaling and hot fluids at a nearby pilot geothermal power plant confirm a meteoric origin. It indicates the current orientations of the main conduits for geothermal fluids are oriented at N10°E with a dip of 70°E [61].

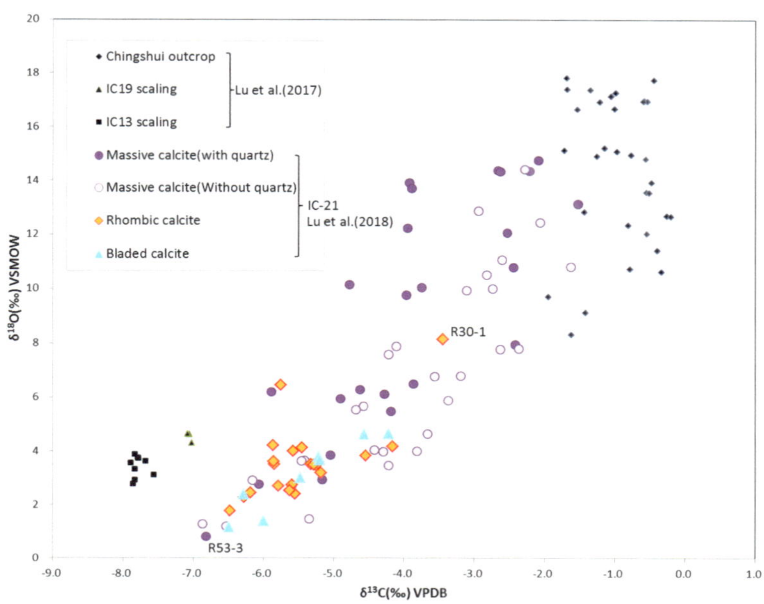

Figure 9.
Plots of carbon and oxygen isotope values of calcite veins and scaling from outcrops, IC-21, and wells in the Chingshui geothermal field.

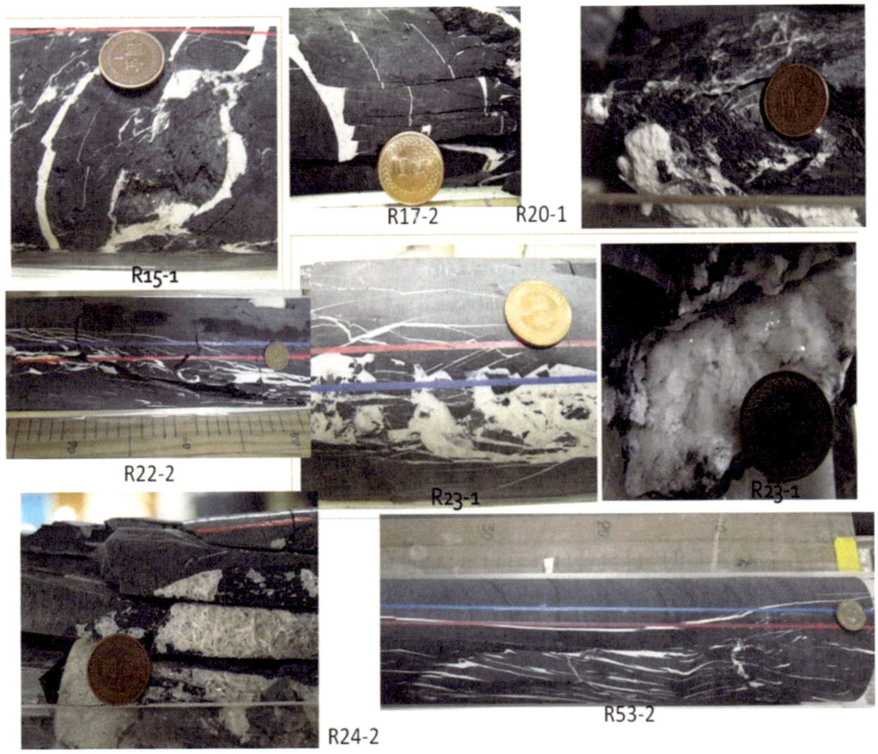

Figure 10.
Photos of veins in the core of well IC-21.

Figure 11.
Photographs of the calcite morphologies observed in the veins of IC-21 cores: massive (yellow arrow), rhombic (red arrow), and bladed calcites (blue arrow) in fractures. The diameter of 1 dollar coin is 2 cm for scales.

6. Conclusions

The Chingshui geothermal field, a moderate-temperature and water-dominated hydrothermal system, was the site of the first geothermal power plant in Taiwan. This article introduces the exploration results of a geothermal reservoir located in the slate formation of Taiwan using geological, geophysical, and geochemical methods. The hot springs, gas fumaroles with sulfur precipitation, hydrothermal alteration, and high surface temperature on rock body are the thermal manifestations in the Chingshui geothermal field. All of these manifestations are predominantly occurring in very narrow belt about 20–30 m wide along the Chingshui River, which are controlled by the fractures of faults. Therefore, surface geological works provide abundant information on rock formations and fracture patterns to infer where and how deep the geothermal reservoir and probably current conduct circulation system are. Furthermore, to construct a model for understanding the subsurface rock bodies and geological structures.

Geophysical surveys including the geomagnetic, gravity, resistivity measurements (transient electromagnetic [TEM] and magnetotelluric [MT]), and microseismicity have been applied to detect the subsurface geological structures. A model with two geothermal reservoirs has been proposed underneath the Chingshui geothermal field. One is shallower at a depth of less than 3000 m, while the other is deeper at depths ranging from 4000 to 8000 m. Moreover, abundant microseismicity distributed at the top of the deep reservoir to infer a high-temperature hydrothermal system with frequent hydraulic fracturing occurred that induces microseismicity. However, the resolutions of geophysical surveys are not so high to precisely draw the whole picture of reservoirs and to pinpoint where the fault fractures as conduits for thermal fluid circulation due to the narrow of fault zones in the slate formation.

Chemical constituents of the Chingshui geothermal water are rich in bicarbonate and sodium in anion and cation, suggesting that it is the HCO_3^--Na^+ fluid type based on Piper diagram. Based on the Na-K-Ca and silica geothermometers, the temperatures of thermal fluids in reservoirs range from 137 to 205°C and from 127 to 214°C, respectively. The H^- and O^- isotopic compositions of thermal water are close relationship with the meteoric water that indicate that the isotopic changes of geothermal water due to fluid–rock interaction were small. This small shift may reflect the slow fluid–rock interaction in terms of low permeability of the slate host rocks.

Carbon and oxygen isotopic analyses indicate that the samples from outcrops and scaling in geothermal wells possess the highest and lowest values, respectively. These results infer that the former could be derived from fluids originating from the shallower reservoir, while the latter may be from the deeper reservoir to precipitate calcite veins near the faults. The calculated oxygen isotopes of fluids combining with the ratios of $^3He/^4He$ suggest that the fluid in the deep may be from magmatic source underneath the Chingshui geothermal field, while the thermal water in shallower reservoir is a mixing between deep magmatic fluids with meteoric one. The original fluids of bladed calcites confirm a meteoric origin, which have the similar oxygen isotopic value with the thermal fluids of Chingshui. It indicates the current orientations of the main conduits for geothermal fluids are oriented at N10°E with a dip of 70°E.

Acknowledgements

This work was supported by the Ministry of Science and Technology (MOST), Taiwan, under the grants of NSC 102-3113-P-002-031, MOST 103-3113-M-002-001, MOST 104-3113-M-002-001, and MOST 105-3113-M-002-001.

Author details

Sheng-Rong Song[1*] and Yi-Chia Lu[1,2]

1 Department of Geosciences, National Taiwan University, Taipei, Taiwan

2 Department of Earth Sciences, National Taiwan Normal University, Taipei, Taiwan

*Address all correspondence to: srsong@ntu.edu.tw

References

[1] BOE. 2016. https://www.moeaboe.gov.tw/ECW/populace/content/ContentLink.aspx?menu_id=137&sub_menu_id=358 [WWW Document]. Bur. Energy, R.O.C

[2] Taipower. No Title [WWW Document]; 2016

[3] Ho CS. A synthesis of geologic evolution of Taiwan. Tectonophysics. 1986;**125**:1-16

[4] Song SR, Yang TF, Yeh YH, Tsao S, Lo HJ. Tatun volcano group is active or extinct? Journal of the Geological Society of China. 2000;**43**:521-534

[5] Moeck IS. Catalog of geothermal play types based on geologic controls. Renewable and Sustainable Energy Reviews. 2014;**37**:867-882. DOI: 10.1016/j.rser.2014.05.032

[6] Song SR. The Geothermal Potentials and Opportunity in Taiwan. IGA-Asian Symposium Geothermal Energy. Offenburg, Germany; 2018

[7] MRSO. Exploration of geothermal resources in Taiwan. Repub. China Rep 2. 1978. 75 p

[8] Song SR. National Energy Program: The Study of Chingshui Geothermal Field (3/3). National Science Council 2012-Final Report (NSC 101-3113-M-002-001); 2012

[9] Bertani R. Geothermal power generation in the world 2010-2014 update report. In: Proceedings of World Geothermal Congress 2015; Melbourne, Australia. 2015. pp. 1-19

[10] Seno T, Maruyama S. Paleogeographic reconstruction and origin of the Philippine sea. Tectonophysics. 1984;**102**: 53-84

[11] Tsai YB, Liaw ZS, Lee TQ, Lin MT, Yeh ZH. Seismological evidence of an active plate boundary in the Taiwan area. Memoir of the Geological Society of China. 1981;**4**:143-154

[12] Yu SB, Chen HY. Global positioning system measurements of crystal deformation in the Taiwan arc-continent collision zone. Terrestrial, Atmospheric and Oceanic Sciences. 1994;**5**:477-498

[13] Ching KE, Hsieh ML, Johnson KM, Chen KH, Rau RJ, Yang M. Modern vertical deformation rates and mountain building in Taiwan from precise leveling and continuous GPS observations, 2000-2008. Journal of Geophysical Research. 2011;**116**. DOI: 10.1029/2011JB008242

[14] Teng LS. Geotectonic evolution of late Cenozoic arc-continent collision in Taiwan. Tectonophysics. 1990;**183**:57-76

[15] Teng LS. Stratigraphic records of the late Cenozoic Penglai orogeny of Taiwan. Acta Geologica Taiwanica. 1987;**25**:205-224

[16] Chen HF. Crustal uplift and subsidence in Taiwan: An account based upon retriangulation results. Spec. Publ. Cent. Geol. Surv. Taiwan. 1984;**3**:127-140

[17] Peng TH, Li YH, Wu FT. Tectonic uplift rates of the Taiwan Island since the early Holocene. Memoir of the Geological Society of China. 1977;2:57-69

[18] Yu SB, Chen HY, Kuo LC. Velocity field of GPS stations in the Taiwan area. Tectonophysics. 1997;**274**:41-59

[19] Yu SB, Kuo LC, Punongbayan RS, Ramos EG. GPS observation of crustal deformation in the Taiwan-Luzon

region. Geophysical Research Letters. 1999;**26**:923-926

[20] Furukawa M, Tokuyama H, Abe S, Nishizawa A, Kinoshita H. Report on DELP 1988 cruises in the Okinawa trough, 2, Seismic reflection studies in the southwestern part of the Okinawa trough. Earthquake Research Institute, University of Tokyo. 1991; **66**:17-36

[21] Kimura M. Back-arc rifting in the Okinawa trough. Marine and Petroleum Geology. 1985;**2**:222-240. DOI: 10.1016/0264-8172(85)90012-1

[22] Sibuet JC, Deffontaines B, Hsu SK, Thareau N, Formal TPL, Liu CS, et al. Okinawa trough backarc basin: Early tectonic and magmatic evolution. Journal of Geophysical Research. 1998;**103**:245-267. DOI: 10.1029/98JB01823

[23] Chen YG, Wu WS, Chen CH, Liu TK. A date for volcanic eruption inferred from a siltstone xenolith. Quaternary Science Reviews. 2001;**20**:869-873

[24] Chang LS. Some planktonic foraminifera from the Suo and Urai Groups of Taiwan and their stratigraphical significance. Proceedings of the Geological Society of China. 1962;**5**:127-133

[25] Chang LS. Tertiary planktonic foraminiferal zones of Taiwan and overseas correlation. Memoir of the Geological Society of China. 1962;**1**:107-112

[26] Chen WS. An introduction to the geology of Taiwan explanatory text of the geologic map of Taiwan (1/400000). Taipei: Geological Society; 2016

[27] Hsiao PT, Chiang SC. Geology and geothermal system of the Chingshui-Tuchang geothermal area, Ilan, Taiwan. Petroleum Geology of Taiwan. 1979;**16**:205-213

[28] Tseng CS. Geology and geothermal occurrence of the Chingshui and Tuchang Districts, Ilan. Petroleum Geology of Taiwan. 1978;**15**:11-23

[29] Lin CW, Lin WH. Explanatory text of the geologic map of Taiwan Sanshin. Sheet 15. Taiwan: Cent. Geol. Surv. MOEA; 1995

[30] Tong LT, Ouyang S, Guo TR, Lee CR, Hu KH, Lee CL. Insight into the geothermal structure in Chingshui, Ilan, Taiwan. Terrestrial, Atmospheric and Oceanic Sciences. 2008;**19**:413-424. DOI: 10.3319/TAO.2008.19.4.000

[31] Yu SB, Tsai YB. Geomagnetic anomalies of the Ilan plain, Taiwan. Petroleum Geology of Taiwan. 1979;**16**:19-27

[32] Su F. Resistivity survey in the Chingshui prospect, I-Lan, Taiwan. Petroleum Geology of Taiwan. 1978;**15**:255-264

[33] Lin CW, Yang CN. Structure styles of the slate and schist belts in northeastern Taiwan. Bulletin of the Geological Survey of Japan. 1999;**12**

[34] Lu YC, Song SR, Wang PL, Wu CC, Mii HS, MacDonald J, et al. Magmatic-like fluid source of the Chingshui geothermal field, NE Taiwan evidenced by carbonate clumped-isotope paleothermometry. Journal of Asian Earth Sciences. 2017;**149**:124-133. DOI: 10.1016/j.jseaes.2017.03.004

[35] Stevenson DJ. Planetary magnetic fields: Achievements and prospects. Space Science Reviews. 2010;**152**:651-664. DOI: 10.1007/s11214-009-9572-z

[36] Lee CR. Compilation of the geothermal prospects data in Taiwan during 1966-1979. Bur. Energy Rep 500; 1994

[37] Keating PB. Weighted Euler deconvolution of gravity data. Geophysics. 1998;**63**:1595-1603

[38] Roy L, Agarwal BNP, Shaw RK. A new concept in Euler deconvolution of isolated gravity anomalies. Geophysical Prospecting. 2000;**48**:559-575

[39] Wightman WE, Jalinoos F, Sirles P, Hanna K. Application of Geophysical Methods to Highway Related Problems. Publication No. FHWA-IF-04-021. Lakewood, CO: Federal Highway Administration, Central Federal Lands Highway Division; 2003

[40] Cheng WT, Lee FC. Bipole-dipole resistivity mapping in the Chingshui geothermal area. Mining and Metallurgy. 1977;**21**:88-93

[41] Chiang SC a, Liu YF. Application of TDEM method in the Chingshui geothermal area, Ilan, Taiwan. Petroleum Geology of Taiwan. 1983;**19**:197-218

[42] Chang PY, Lo W, Song SR, Ho KR, Wu CS, Chen CS, et al. Evaluating the Chingshui geothermal reservoir in northeast Taiwan with a 3D integrated geophysical visualization model. Geothermics. 2014;**50**:91-100. DOI: 10.1016/j.geothermics.2013.09.014

[43] Ho G, Chang P, Lo W, Liu C, Song S. New evidence of regional geological structures inferred from reprocessing and resistivity data interpretation in the Chingshui-Sanshing-Hanchi area of southwestern Ilan County, NE Taiwan. Terrestrial Atmospheric and Oceanic Sciences. 2014;**25**:491-504. DOI: 10.3319/TAO.2014.01.24.01(TT)1

[44] Uchida T, Song Y, Lee TJ, Mitsuhata Y, Lim SKL, Lee SK. Magnetotelluric survey in an extremely noisy environment at the Pohang low-enthalpy geothermal area, Korea. In: Proceedings of World Geothermal Congress. Turkey; 2005

[45] Ushijima K, Mustopa EJ, Jotaki H, Mizunaga H. Magnetotelluric soundings in the Takigami geothermal area, Japan. In: World Geothermal Congress. 24-29; 2005

[46] Wannamaker PE, Rose PE, Doerner WM, Berard BC, McCulloch J, Nurse K. Magnetotelluric surveying and monitoring at the coso geothermal area, California, in support of the enhanced geothermal systems concept: Survey parameters and initial results. In: Proc. World Geotherm. Congr. 2005; April 2005 24-29; Antalya, Turkey. 2004. pp. 24-29

[47] Jones AG, Chave AD, Egbert G, Auld D, Bahr K. A comparison of techniques for magnetotelluric response function estimation. Journal of Geophysical Research - Solid Earth. 1989;**94**:14201-14213. DOI: 10.1029/JB094iB10p14201

[48] Chiang CW, Hsu HL, Chen CC. An investigation of the 3D electrical resistivity structure in the Chingshui Geothermal area, NE Taiwan. Terrestrial, Atmospheric and Oceanic Sciences. 2014;**26**:269-281. DOI: 10.3319/TAO.2014.12.09.01(T)

[49] Liu HF. Study of microseismicity and traveltime tomography in the Chingshui geothermal area [master's thesis]. National Taiwan University; 2013

[50] Mwangi SM. Application of geochemical methods in geothermal exploration in Kenya. Procedia Earth and Planetary Science. 2013;**7**:602-606

[51] Piper AM. A graphic procedure in the geo-chemical interpretation of water analysis. Transactions of the American Geophysical Union. 1953;**25**:914-928. DOI: 10.1029/TR025i006p00914

[52] Huang JS, Peng DH. Subsurface geological report of IC-4 well

Ching-Shui, I-Lan, Taiwan. Pet. Explor. Div. 1976:1-13

[53] Huang YH, Liu HL, Song SR, Chen HF. An ideal geothermometer in slate formation: A case from the Chingshui geothermal field, Taiwan. Geothermics. 2018;**74**:319-326. DOI: 10.1016/j.geothermics.2017.11.002

[54] Liu KK, Yui TF, Shieh YN, Chiang SC, Chen LH, Hu JY. Hydrogen and oxygen isotope compositions of meteoric and thermal waters from the Chingshui geothermal area, northeastern Taiwan. Proceedings of the Geological Society of China. 1990;**33**:143-165

[55] Iwatsuki T, Satake H, Metcalfe R, Yoshida H, Hama K. Isotopic and morphological features of fracture calcite from granitic rocks of the Tono area, Japan: A promising palaeohydrogeological tool. Applied Geochemistry. 2002;**17**:1241-1257. DOI: 10.1016/S0883-2927(01)00129-9

[56] Li R, Dong S, Lehrmann D, Duan L. Tectonically driven organic fluid migration in the Dabashan Foreland Belt: Evidenced by geochemistry and geothermometry of vein-filling fibrous calcite with organic inclusions. Journal of Asian Earth Sciences. 2013;**75**: 202-212. DOI: 10.1016/j.jseaes.2013.07.026

[57] Luetkemeyer PB, Kirschner D, Huntington KW, Chester JS, Chester FM, Evans JP. Constraints on paleofluid sources using the clumped-isotope thermometry of carbonate veins from the SAFOD (San Andreas Fault Observatory at Depth) borehole. Tectonophysics. 2016. DOI: 10.1016/j.tecto.2016.05.024

[58] Wallin B, Peterman Z. Calcite fracture® fillings as indicators of paleohydrology at Laxemar at the Aspo Hard Rock Laboratory; southern Sweden. Applied Geochemistry. 1999;**14**

[59] Wang PL, Wu JJ, Yeh EC, Song SR, Chen YG, Lin LH. Isotopic constraints of vein carbonates on fluid sources and processes associated with the ongoing brittle deformation within the accretionary wedge of Taiwan. Terra Nova. 2010;**22**:251-256. DOI: 10.1111/j.1365-3121.2010.00940.x

[60] Cheng Y. Geochemical characteristics of groundwater in the Ilan Plain, Northeast Taiwan [master's thesis]. Natl. Taiwan Univ. 2014. pp. 1-82

[61] Lu YC, Song SR, Taguchi S, Wang PL, Yeh EC, Lin YJ, et al. Evolution of hot fluids in the Chingshui geothermal field inferred from crystal morphology and geochemical vein data. Geothermics. 2018;**74**:305-318. DOI: 10.1016/j.geothermics.2017.11.016

[62] Domingo C, Loste E, Garc J, Fraile J. Calcite precipitation by a high-pressure CO_2 carbonation route. The Journal of Supercritical Fluids. 2006;**36**:202-215. DOI: 10.1016/j.supflu.2005.06.006

[63] Lahann RW. A chemical model for calcite crystal growth and morphology control. Journal of Sedimentary Research. 1978;**48**:337-347

[64] Moore CH, Wade WJ. Carbonate diagenesis: Introduction and tools. In: Carbonate Reservoirs. Elsevier Inc; 2013. pp. 1-392

Chapter 3

Geothermal Potential of the Global Oil Industry

Jon Gluyas, Alison Auld, Charlotte Adams, Catherine Hirst, Simon Hogg and Jonathan Craig

Abstract

There are around 40 new geothermal power projects commissioned in each of the last few years. Growth of the market is around 5% annually and current installed capacity is about 13,300 MW with about the same in development in 24 countries. These figures are impressive, but they do not bear comparison with any of the fossil fuels. However, few will realise that the global oil industry has a cryptic geothermal power potential that is equal to the entire current output of the geothermal industry. The oil industry is ageing. Many areas still produce copious quantities of oil, but the oil comes with an unwanted by-product, water. The volume of water produced is typically is 10–20 times that of the oil; and the water is hot—in some places very hot (>100°C). In a recent study we showed that the power depleted oil production platforms of the North Sea's North Viking Graben produce sufficient hot water to deliver around 60% of the power requirement for each field. A review of global oil and hence water production has enabled us to calculate that power production alone from waste water from producing oilfields could be at least 15,000 MW.

Keywords: geothermal energy, waste water, power production, organic-rankine engine, geothermal heat

1. Introduction

In 2014, power production across the whole of the global geothermal industry was around 12,000 MW [1]. Once in production, geothermal power plants have a near zero carbon footprint and because of the constancy of the temperature of the earth at depth they tend to provide a good base load power supply.

The oil industry also produces hot water, some of which could be used to generate power either to supplement oilfield operations or for export. In particular as oil fields age there is a tendency for water (co) production to increase as the oil becomes depleted and because in most instances the oil is more viscous than the water in associated aquifers. In consequence water can become the most dominant fluid produced. This is particularly the case where water is injected on the periphery of a field or beneath the oil zone. The injected water can re-emerge at the production wells. Such water is heated by its passage through the matrix of the reservoir and in many instances the produced water is at the same or nearly the same temperature as the oil.

Hot water can be employed to generate power using an organic Rankine cycle engine. Here we examine global co-produced water production of the oil industry and hence the potential for geothermal power generation.

2. Global oil production and co-produced water

Global oil production stood at 87 million barrels per day in 2013, having risen steadily for over 25 years. Even the global economic downturn at the end of the last decade had only a small impact on this trend, reversing it for 1 year only in 2009 (**Figure 1**). During the past few decades several new oil provinces have come into production, (e.g. West Africa, sub-salt Brazilian Atlantic Margin and deep water Gulf of Mexico) but more than 85% of current global production still comes from old established provinces such as those in the Middle East, Russian Federation, China, Canada, Venezuela and the North Sea. All of which have been in production for tens of years. The major fields in the North Sea have been on production for 40 years; those in Saudi Arabia, Kuwait and Venezuela around 80 years, while fields in the Zagros Province of Iran and parts of the Russia are well past their first century of production.

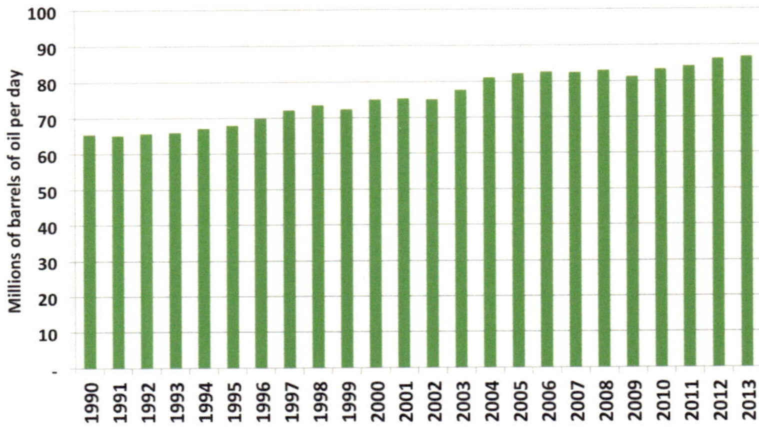

Figure 1.
Global oil production from 1990 to 2013 [2].

One thing that all of these old provinces have in common is that as well as producing oil they also produce water; water in copious amounts. Most ageing fields display a growing water production trend. The produced water may be from naturally inflowing connate (formation) water and/or water that has been injected to maintain reservoir pressure and promote sweep of the oil from the reservoir to the production wells. There are some fields that produce little water throughout their life but, typically, only a small proportion of the oil in these fields is produced (5–20%), mainly by gas exsolution drive. Indeed, it is common practice in such fields for water injection to be installed to increase the oil recovery. Much of this injected water eventually makes its way to the production wells.

The viscosity of most oils is higher than that of water, so the flow to the production well is promoted over that of water. In mature fields with active aquifers, or those that have pressure support from water injection, water will become an ever increasing and unwanted production byproduct. Water oil ratios in mature fields are typically in the 10–20 range.

3. A note on units

The petroleum industry typically follows US practice and measures production in barrels (bbl) per day. About 6.29 barrels are equivalent to 1 m^3 of water. One cubic meter of water weighs 1 tonne exactly, if fresh, or up to about 1.1 tonnes if brine, depending on the solution strength. In Europe, national bodies typically report production in partially converted SI units, with the added complication that the UK Department of Energy and Climate Change reports in thousands of cubic meters per month.

The geothermal industry tends to report flow rates in litres per second. Thus.

$$1\,\mathrm{l\,s^{-1}} = 86{,}400\ \mathrm{l\ per\ day} = \text{about 543 barrels per day.} \quad (1)$$

This simple difference in units used means that oil and geothermal industry production figures are rarely compared, so the importance of the oil industry's geothermal water resource has not been widely appreciated.

4. Examples of co-produced water N Sea fields

The North Sea emerged as a globally significant oil province in the 1970s. Many giant fields were discovered in quick succession as exploration that had begun between 56° and 58° N in both the UK [3] and Norwegian sectors spread northwards and into Arctic waters (for Norway). By 1975 the first field, Argyll was on production [4] and a year later the first of the giant fields, Forties, was on production [5]. Most of the fields discovered in the North Sea are still on production, although this is changing following the fall of the oil price during the latter half of 2014 and into 2015. Nonetheless, even those fields that remain on stream will be well past their prime, with oil production levels maybe only 5 or 10% of peak levels. The modest flow of oil is accompanied by large quantities of water, typically 10–20 times greater than the oil volume. The water is a mix of injected (sea) water and connate water. Water injection is used to maintain reservoir pressure and so support production. The injected water sweeps the oil away from the injection sites towards the production wells. In many fields, the water handling is a 'bottleneck', the volume being at the limit of the handling and cleanup capacity of the facilities prior to reinjection and discharge.

For example, the Murchison Field (**Figure 2**), located in the North Sea's Brent Province, produces oil from high-quality, Middle Jurassic paralic sandstones. It came on stream in 1979 and for the first 3 years of production it produced little water (brine). Plateau oil production for the field was maintained at about

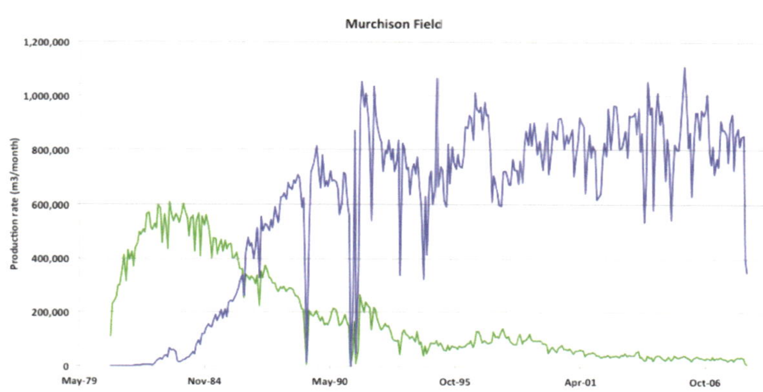

Figure 2.
Oil and water production profile for the Murchison field, North Sea.

600,000 m^3/month until late 1984, by which time water production had risen from near zero to 200,000 m^3/month. Total fluid production of around 800,000 m^3/month was maintained from 1985 until at least 2008 but, over time, the ratio of water to oil has risen to about 30:1. The Murchison Field is located at about 3 km depth and produces oil and water at 110°C. After cleaning, the produced water is discharged to the sea [6].

Murchison is typical of the ageing suit of North Sea Fields. Many are now producing at least 10x, and some 20x, as much water as oil. While the oil price remained high relative to operating costs, production continued, but with the sustained low oil prices since mid-2014 operators are now opting to abandon fields. Water handling (processing and cleaning) is a major cost and can limit the viability of the field, even though only half the oil in the field may have been recovered. Thus, the hot water resource in the North Sea fields is disappearing fast, as it will in other oil producing regions as fields are abandoned. The situation might be different if instead of being a resource the hot water was instead considered a reserve.

5. Southampton district energy scheme & Wytch farm

The UK has a single producing geothermal well at Southampton in the Wessex Basin [7]. This well was one of six drilled in the 1980s on behalf of the British government. They were planned in the wake of the first oil crisis of the 1970, and before the UK became a significant petroleum producing country, in order to evaluate the geothermal potential of the UK. By the time the wells were drilled, the UK was a global leader in oil and gas production from the North Sea and this, coupled with difficulties encountered in retrieving any hot water from the wells drilled into the relatively hot Cornish granites, meant that the geothermal aspirations for the UK remained unfulfilled. Michael Smith, the accountant at Southampton District Council, however, recognised the potential of a zero cost, 2 km hole in the ground. Hot water produced by this well became the locus for the development of a combined heat and power project—the Southampton District Energy Scheme.

The well at Southampton was drilled in 1981. It encountered the Triassic Sherwood sandstone at a depth of between 1730 and 1800 m below mean sea level (**Figure 3**). The aquifer was 70 m thick where penetrated and 24 m of this was deemed producible. The measured temperature was 76°C and flow rate of the well was 10–15 l s^{-1}. Construction of the facility began in 1987 and, at the time, it was estimated to have a 10–15 year life [7]. In fact, it continues to operate today, albeit at reduced flow rate, and the thermal output is about 1.7 MW (at 2 l s^{-1}). The 'heat-depleted' water is not reinjected into the reservoir but is discharged at about 50°C into the nearby Southampton Water (sea).

Figure 3.
Triassic Sherwood sandstone exposed in the coastal cliffs of Ladram Bay, Devon, UK. This sandstone has high net to gross, high porosity, high permeability and a high vertical to horizontal permeability ratio. It also forms the main producing reservoir in the Wytch farm oilfield and the producing horizon in the Southampton District energy scheme, both located further east along England's south coast (photograph by J. Gluyas).

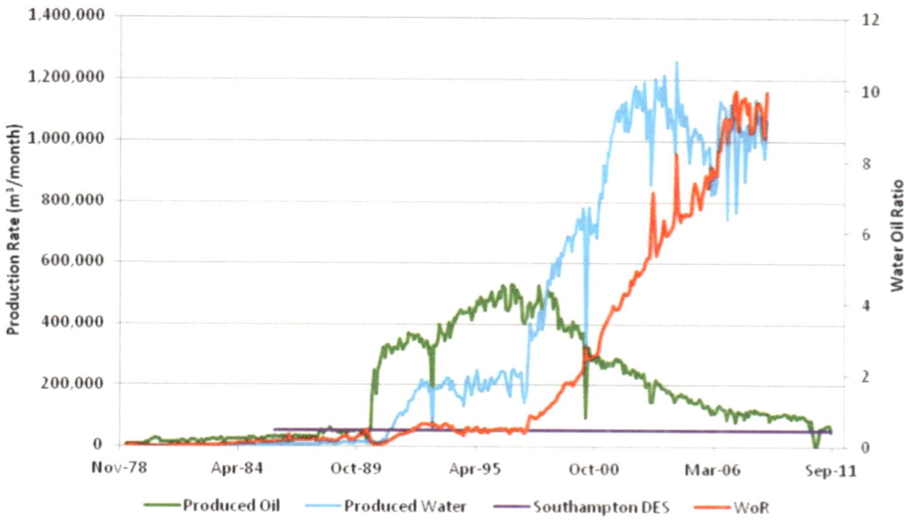

Figure 4.
Comparison of production of water from the Southampton District energy scheme (purple) with production from Wytch farm (oil = green, water = blue and the water to oil ratio = red). At peak, Wytch farm was producing 40x as much fluid as Southampton District energy scheme.

Aside from the fact that the Southampton District Energy Scheme is the only geothermal project in the UK, there is nothing remarkable about it. What is remarkable is that only a few tens of kilometres away from Southampton another 'scheme' was producing 40 times as much hot (67°) water, none of which has been used for any district heating.

The Wytch Farm Field in Dorset is the UK's largest and Europe's second largest onshore oilfield (**Figure 4**). Developed initially in the 1970s, it was later extended in the 1990s to exploit the larger and deeper Triassic Sherwood Sandstone reservoir. Petroleum production peaked in the mid-1990s at 0.5 million m^3/month while, 10

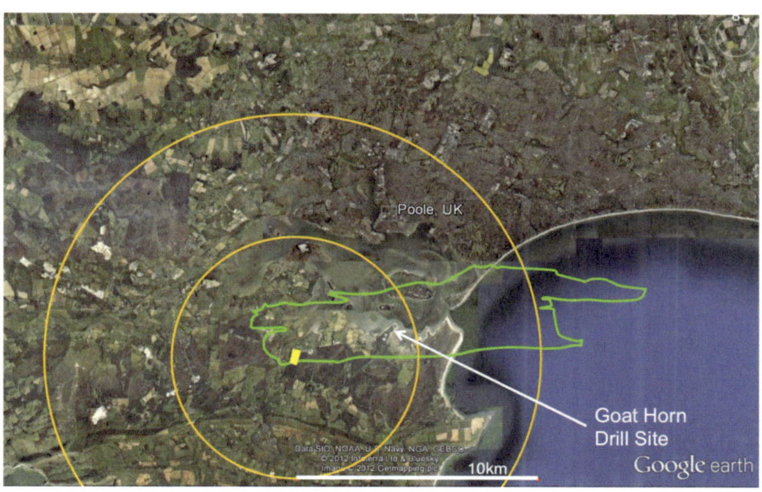

Figure 5.
Location of the Wytch farm oilfield (outline in green) relative to the Poole and Bournemouth (top right hand corner of figure) urban areas. The concentric circles are 5 and 10 km radius and centred on the Wytch farm gathering station. District heating schemes typically lose up to about 1°C km^{-1} of insulated pipe [8]. GH is the location of the Goathorn peninsula from where long reach, horizontal wells were drilled into Poole harbour and is the location of the well heads closest to Poole-Bournemouth [9]. Base image from Google earth.

years later, total fluids production reached 1.5 million m^3/month, 90% of which was water. Oil is exported from the gathering station via pipeline and, during the build up to peak oil production, a chiller was installed so that the exported oil did not warm the earth surrounding the pipeline. The produced water is either cleaned and disposed of, or is re-injected for water maintenance. At peak, the field was producing around 570 l s^{-1}, compared with the 10–15 l s^{-1} from the Southampton well. Hence, the flow rates and hence low enthalpy geothermal potential of the Wytch Farm Field exceeded that of the Southampton project by some fortyfold.

Despite the proximity of Wytch Farm to the urban areas of Poole and Bournemouth (population c. 350,000 in 2015; **Figure 5**), no one seems to have considered the possibility that the co-produced (and hence 'free') geothermal water from Wytch Farm could be used—indeed could still be used—to heat commercial and domestic property in these areas.

6. Power generation

For most of the geothermal energy industry, dry steam is used to drive steam turbines and so generate electricity. The steam is produced by circulating water through hot rocks close to the Earth's surface. Temperatures typically need to be 150°C or greater. However, lower temperature wet steam and even hot water can be used to generate electricity using a variety of heat engines, the most common of which are organic Rankine cycle engines (ORC). In these, the hot water is used to vaporise a volatile fluid and the vapour drives the turbine. The vapour is then condensed and the cycle begun again. For example, at Chena Hot Springs in Alaska, USA, hot water at about 73°C is used to generate around 400 kW of geothermal power using ORC [10]. The critical factor is the difference between the temperature of the hot water and that of the condensing fluid, not the absolute temperature of the water.

In a recent study, Auld et al. [11] examined the power production potential for a suite of oilfields in the Northern North Sea (Brent Province), including the aforementioned Murchison Field. Here, the condensing fluid is the cold North Sea (approx. 5°C). For many fields in an offshore setting, the co-produced gas that exsolves from the oil as it is brought to surface, is used to produce electrical power. Gas turbines generate the power required to run the platform. The largest consumer of power are, typically, the large water pumps used to inject water into the reservoir and so maintain pressure and assist in the sweep of oil to production wells. In this regard, the Brent Province is problematic. Its oils contain relatively little dissolved gas and as the oil production declines so does the gas production. For many fields, this reduction in gas supply means that the fields cannot be run optimally, as not enough water can be injected to maintain commercial rates of oil production. For one company at least, failure to produce oil at commercial rates has led to the premature abandonment of a field with only 45% of the in place oil recovered. In mid-2015, Fairfield Energy announced their plans to abandon the Dunlin Field and its satellites [12]. The abandonment may have been postponed had the co-produced water from the field been used to produce power and offset operational costs as well as, in this instance, the cost of building of a new, expensive gas import pipeline that was little used before the abandonment of the field was announced.

Auld et al. [11] demonstrated that the six Brent Fields studied in detail could have produced more than 10 MW and one, Ninian, 31 MW (**Figure 6**). A comparison of the cost of the ORC system sized for 10 MW power generation showed that payback was between 3 and 4.5 years. This payback time did not account for

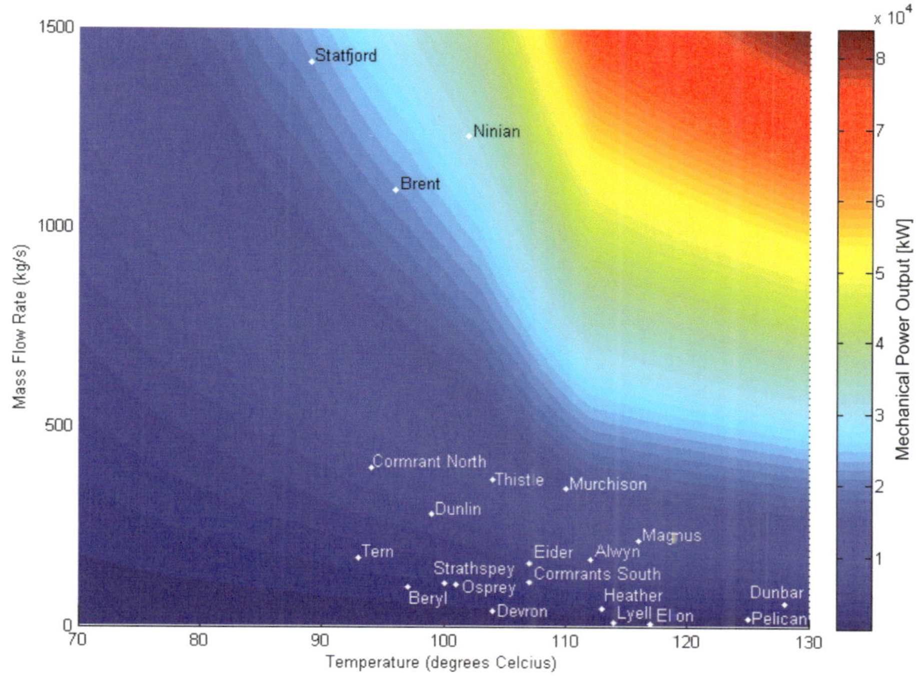

Figure 6.
Brent Province oilfields, potential power output of organic Rankine cycle fuelled by co-produced hot brine (from [11]).

greenhouse gas emission levies that would shorten payback, nor for the additional benefit of using the spent cold brine for reinjection into the reservoir. The increased viscosity of the cooler injection water improves its mobility ratio compared with that of oil and hence improves sweep efficiency.

7. UK co-produced water

The UK Department of Energy and Climate Change continuously releases near up to date oil and water production figures and water injection figures for all UK offshore oilfields. The North Sea production peaked in 2000. At peak about 8 million barrels of oil plus water were being produced every day, of which about 5 million barrels were water. Since 2000, production has declined with oil production falling faster than water production. Indeed, water production was still 5 million barrels per day in 2007, but it has fallen somewhat since (**Figure 7**).

Many of the North Sea fields produce from depths of around about 3 km and, although there is some variation, the geothermal gradient over much of the North Sea is about 30°C km^{-1}, a little higher around the triple (rift) junction in the Central North Sea. Using field and reservoir temperature data published in [13] for 69 fields (data tables at end of each chapter), the average temperature of oil and brine produced form North Sea Fields is 108 ± 26°C (**Figure 8**). Of these 69 oil fields, over 80% are at temperatures in excess of 90°C. Using the approach of [11], a daily production of water of 5 million barrels, a reservoir temperature of 100°C and a condensation temperature of 5°C yields a theoretical power resource, using all available co-produced water to generate power, for the North Sea of around 250 MW.

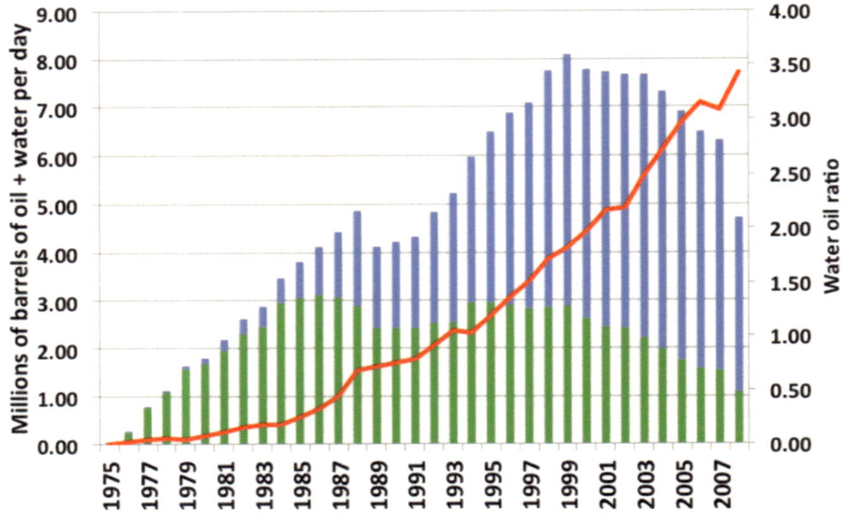

Figure 7.
UK North sea oil production (green) and co-produced water (blue) with water oil ratio trend in red.

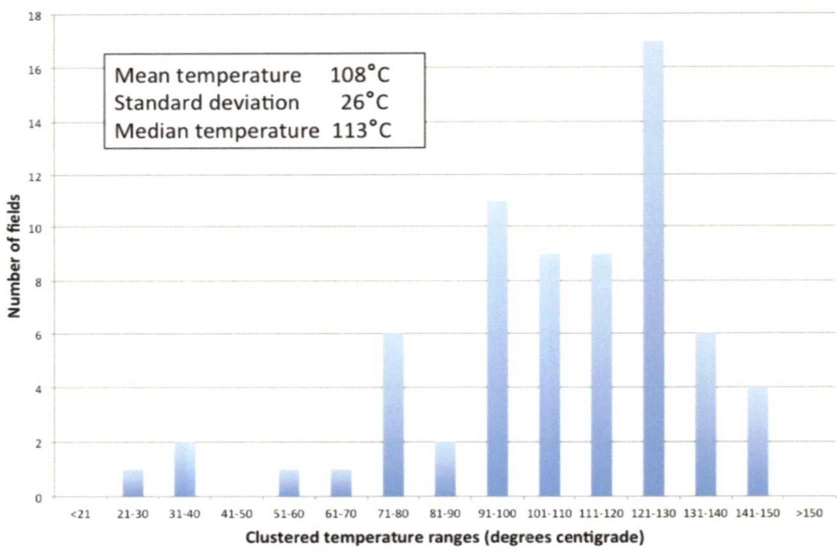

Figure 8.
Temperature distribution of North Sea oilfields (from [12]).

8. Global co-produced water

Translating the geothermal power potential of the North Sea fields to make an estimate for the whole world is extremely difficult; not because of the calculation itself is difficult, but because data on temperature of producing fields and volumes of co-produced water are kept confidential by many companies and countries. This is particularly true for the water production data, because it is straightforward to use the evolving water to oil ratio in a field, or indeed basin, to estimate remaining reserves and remaining reserves are often state secrets and the data that are release are often subject to political overprint. Nonetheless, it is possible to make a broad estimate of global water production, particularly in view of the known and changing discovery rate for petroleum over the decades (peak finding time was the 1960s).

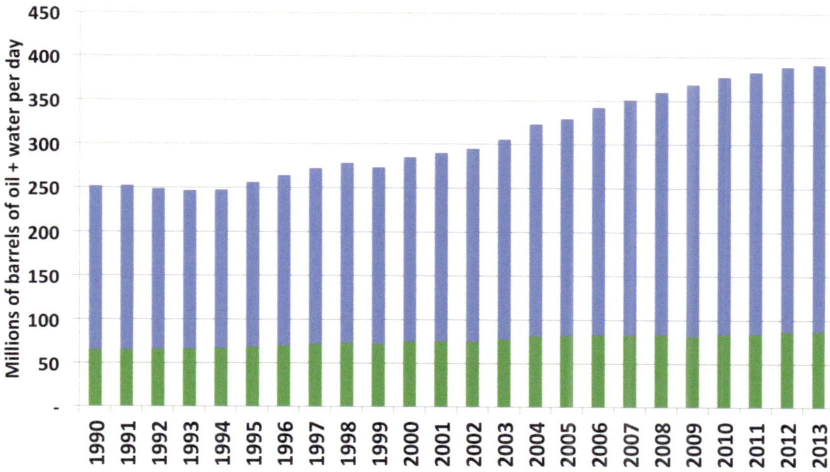

Figure 9.
Global water production amounts to about 300 million barrels per day from which as much as 15,000 MW of power could be generated.

To do this, we have used a combination of primary and secondary sources, most notably: Dal Ferro and Smith [14] and an anonymous white paper from the Society of Petroleum Engineers [15, 16] on 'Challenges in Reusing Produced Water'. **Figure 9** combines global oil production data from the aforementioned BP Statistical Review of World Energy and the references cited above. Global water production is around about 300 million barrels per day and about 3.5 times as much as daily oil production (**Figure 9**). About 70% of oil (and water) production is from onshore. We do not know the average temperature of oilfields around the world but speculating that it is not much lower than that of the North Sea would lead to the conclusion that waste water from the oil industry could deliver around about 15,000 MW. The very large assumptions used to make this calculation mean that this figure should be treated with some caution. Nonetheless, such a figure is about the same as the total current global output of the geothermal industry. This implies that the untapped geothermal resource intrinsic within the oil industry could, if used, double geothermal power production across the globe and at modest cost since the hot water is already produced. In addition to the potential power generated the substantial residual thermal energy could also be used in many parts of the world.

Moreover, as oilfields age further, efforts to maintain oil production are likely to result in an ever-increasing water oil ratio and, most likely, the production of more water – and, hence, more geothermal resource.

9. Conclusions

Today, dedicated geothermal plants produce about 13 MW globally. The oil industry could generate much the same amount of power by processing the co-produced water from ageing fields to produce power using Organic Rankine Cycle engines. If co-produced water was considered as an asset rather than a waste product, it is likely that oil fields would be operated longer with ever-increasing water oil ratios and the geothermal utility of the fields would increase. Utilisation of the geothermal resources from oilfields would also help curb emissions by reducing the amount of co-produced gas used for power generation and ultimately allow more oil to be recovered from the fields because they could be run economically for longer with even lower oil production rates.

Author details

Jon Gluyas[1*], Alison Auld[2], Charlotte Adams[3], Catherine Hirst[4], Simon Hogg[5] and Jonathan Craig[6]

1 Department of Earth Sciences, Durham University, Durham, UK

2 Diagio Plc, Moray House, UK

3 Department of Geography, Durham University, Durham, UK

4 COWI UK Ltd, Glasgow, UK

5 School of Engineering and Computing Sciences, Durham University, Durham, UK

6 Eni Upstream and Technical Services, Milan, Italy

*Address all correspondence to: j.g.gluyas@durham.ac.uk

References

[1] Geothermal Energy Association. Annual US & Global Geothermal Power Production Report, Geothermal Energy Association. April 2014. Available form: http://geo-energy.org/events/2014%20Annual%20US%20&%20Global%20Geothermal%20Power%20Production%20Report%20Final.pdf [Accessed: July 25, 2017]

[2] BP. BP Statistical Review of World Energy. 2014. Available from: https://www.bp.com/content/dam/bp-country/de_de/PDFs/brochures/BP-statistical-review-of-world-energy-2014-full-report.pdf [Accessed: September 11, 2018]

[3] Hardman RFP. Lessons from oil and gas exploration in and around Britain. In: Gluyas J, Hichens HM, editors. United Kingdom Oil and Gas Fields–Commemorative Millennium. UK: Geological Society of London (Memoir No. 20); 2003. pp. 5-16

[4] Gluyas JG, Mair B, Schofield P, Arkley P, McRae D. Ardmore field: Rebirth of the first offshore field UKCS. In: Dore AG, Vining BA, editors. Petroleum Geology: North-West Europe and Global Perspectives, Proceedings of the Sixth Petroleum Geology Conference. London: Geological Society; 2005. pp. 367-388

[5] Wills JM. The Forties Field, Blocks 21/10, 22/6a, UK North Sea. In: Abbots IL, editor. United Kingdom Oil and Gas Fields, 25 Years Commemorative Volume. UK: Geological Society (Memoir 20); 1991

[6] Ray JP, Engelhardt FR. Produced Water, Technological/Environmental Issues and Solutions, Environmental Research Volume 46. Springer; 1992. 616 p

[7] Gearty M, Clark W, Smith M. Southampton District energy scheme, a story of collaboration and stead ambition, A learning history. Issue VWS workshop, Number 5. 2008. Available from: http://people.bath.ac.uk/mnspwr/doc_theses_links/pdf/dt_mg_APPDXGSouthampton_case_issue_vws_A5.pdf [Accessed: July 25, 2017]

[8] Chilton T, Forsyth I. Leicester District Energy Scheme. Cofely: District Energy Limited; 2012. Avaialble from: http://www.climate-em.org.uk/images/uploads/EM_masterclass_200112.pdf [Accessed: July 25, 2017]

[9] Gluyas JG, Swarbrick RE. Petroleum Geoscience. Blackwells; 2004. 359 p

[10] Holdmann G. The Chena Hot Springs 400 kW geothermal power plant: Experience gained during the first year of operation. In: Chena Geothermal Power Plant Report. Alaska: Chena Power Plant; 2006. pp. 1-9

[11] Auld A, Hogg S, Berson A, Gluyas JG. Power production via North Sea hot brines. Energy. 2014;**78**:674-684

[12] Gluyas JG, Swarbrick RE. Petroleum Geoscience. 2nd ed. Wiley; 2019

[13] Gluyas JG, Hichens H. North Sea Oil and Gas Fields-Millennium Review. UK: Geological Society of London (Memoir 20); 2003. p. 1006

[14] Dal Ferro B, Smith M. Global onshore and offshore water production. In: Oil & Gas Review OTC Edition. 2007. Available from: http://www.touchoilandgas.com/global-onshore-offshore-water-a7137- 1.html

[15] SPE. Challenges in Reusing Produced Water SPE. 2010. Available from: http://www.spe.org/industry/challenges-in-reusing-produced-water.php

[16] Veil JA, Puder MG, Elcock D, Redweik Jr RJ. A white paper describing produced water from production of crude oil, natural gas and coal bed methane. Argonne National Laboratory, Prepared for US Department of Energy, National Energy Technology Laboratory Under Contract W-31-109-Eng-38. 2004

Chapter 4

Stages of a Integrated Geothermal Project

Alfonso Aragón-Aguilar, Georgina Izquierdo-Montalvo, Daniel Octavio Aragón-Gaspar and Denise N. Barreto-Rivera

Abstract

A geothermal project constitutes two big stages: the exploration and the exploitation. Each one has a single task whose results allow defining the feasibility of a geothermal project, until achieving the construction and operation stage of the power generation plant. The first stage contains the area recognition, its limitation to the target, and elimination of external factors until defining a geothermal zone with characteristics to be commercially exploited. The main studies and analysis that can be applied during the exploration stage are listed, and the major indicator to continue with the project or suspend is the prefeasibility report. The major risks in the exploration stage are due to studies that are carried out on the surface; at this stage, the costs can be considered low. The main results of the exploration are the selection of sites to drill three or four initial wells. Each well provides a direct overview of the reservoir: depth, production thicknesses, thermodynamic parameters, and production characteristics. The drilling of three to four exploratory wells is recommended, as far as there is certainty of the feasibility of the project, and the development of the field begins with drilling of sufficient wells to feed the plant. In this stage, the cost increases, but the risks decrease.

Keywords: geothermal project stages, regional exploration, drilling exploratory, drilling well locations, costs during exploration, producer wells, geothermal direct uses

1. Introduction

The geothermal word is used to refer to the heat existing inside the earth. From a practical viewpoint, geothermal is the study and the use of the heat energy that by conduction through the rock or transported by fluids moves from the interior of the earth's crust to the surface levels to form geothermal reservoirs.

The energy stored in the form of heat in the rock and aquifers located near the surface can be exploited by drilling wells of up to 3000 or 4000 m depth. The geothermal reservoirs are sometimes visualized on the surface in the form of mud volcanoes, fumaroles, geysers, hydrothermal springs, etc.

The meteoric water that infiltrates through permeable rocks at great depths is heated directly or indirectly by the flow of heat generated by the magmatic chambers, reaching high-, medium-, or low-enthalpy geothermal reservoirs. A geothermal field susceptible of exploitation by steam production for the purpose of electric generation or only hot water (low enthalpy) must present the following characteristics:

- A thermal anomaly
- A reservoir constituted by permeable rocks where the geothermal fluid circulates at economically exploitable depths
- An impermeable cover that prevents the loss of heat by circulation of the geothermal fluid to the surface

In the exploration of a geothermal region, to locate a thermal anomaly (heat source) near the earth's surface, volcanological, structural, and petrological methods are applied. These methods help to distinguish volcanic centers within the regional structural framework estimating their age. Morphological, stratigraphic, and radiometric methods are also applied, which allow the geometry of the geological units in the subsoil to be reconstructed in broad strokes.

The elements that favor the existence of geothermal areas are the persistence of volcanic activity and the frequent eruptions of products that need for its formation a long stay of magma in a chamber. In most cases the chambers are large deposits of magma that feed central volcanic complexes.

Most of the geothermal fields are located in areas of volcanism constituted by clearly identified products (andesite, rhyolites, or dacites) related to an igneous intrusion with depth between 10,000 and 15,000 m representing the magmatic chamber of recent or active volcanic centers.

The five main countries of installed capacity for generating electricity, in the world, using geothermal resources [1–3] are shown in **Table 1**. This classification includes the statistics of production in these five countries between 2005 and 2015.

However, recent additions in 2017 show an increase of 792 MW around the world [4]. Such new additions are as follows:

- Turkey: 325 MW.
- Indonesia: 359 MW.
- Chile: 48 MW.
- Iceland: 45 MW.
- Mexico: 25 MW.
- United States: 24 MW.

	Installed capacity (MW)			
	2005	**2010**	**2015**	**Growth rate (MW)**
United States	2564	3098	3450	886
Philippines	1930	1904	1870	−60
Indonesia	797	1197	1340	543
Mexico	953	958	1017	64
New Zealand	791	843	1005	214

Table 1.
Classification of the first five countries with the largest installed capacity for generating electricity, from geothermal resources, in the world, in the period 2005–2015 [1–3].

- Portugal (Azores): 3 MW.
- Japan: 5 MW.
- Hungary: 3 MW.

It can be seen from **Table 1** that in 2005 Mexico was in the third place among the countries with the highest installed capacity for electricity generation; currently in 2015 it is located in the fourth position. In this period the United States of North America had an increase in its installed capacity of 886 MWe, Indonesia with 543 MWe, New Zealand with 214 MWe, and Mexico with 64 MWe. The Philippines showed a decrease of 60 MWe in its installed capacity to generate electricity with geothermal resources.

A geothermal reservoir must contain a volume of fluid large enough to ensure prolonged exploitation to recover the investment made. It must also be located within a hydrological system that allows the hydraulic recharge of the area in operation. When the balance between exploitation and recharge is maintained, it is possible to consider geothermal energy as renewable and sustainable.

2. Stages of a geothermal project

In general, the execution of a typical geothermal project is divided into two main parts: (a) exploration and (b) exploitation. The first is considered high uncertainty whose objective is the reservoir identification, including a study of its possible use. It also implies significant levels of economic risk that must be faced with progressively increasing but relatively low-cost investments. The exploitation stage involves minor risks but requires high investments.

The Latin American Energy Organization (OLADE) [5] drafted a guidance document for recognition studies of a geothermal project, applicable to Latin American countries. The document states that from a practical point of view, a geothermal project can be divided into the stages, appearing in **Figure 1**.

According to the experiences obtained in the world, it has been identified that for areas of research coverage of a thermal source, during the exploration the following ranges can be referred:

For the survey study, the area to be covered could be ≥1000 km^2.

- For the prefeasibility study, the area could have a coverage between 400 and 500 km^2.
- The area for a feasibility study is reduced to between 10 and 100 km^2.

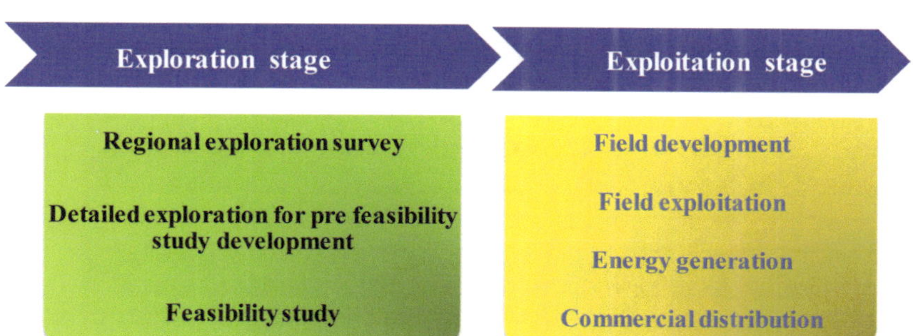

Figure 1.
Two main stages (exploration and exploitation) of a geothermal project described by [5], with its corresponding particular tasks.

To develop a project in a geographically understudied region, it is necessary to start the exploration with a reconnaissance study that covers an area of 1000 km^2 or more. This recognition will make it possible to formulate the first hypotheses about the geothermal possibilities of the region and select one or several favorable areas to carry out prefeasibility studies. Likewise, a detailed exploration program applicable to areas of geothermal interest will be designed.

The objective of prefeasibility tasks is to identify, with field studies, the possible existence of underground reservoir under conditions such that if the process continues with deep perforations, the risk will be minimal. Based on the survey area, prefeasibility investigations are usually carried out in areas ranging from 400 to 500 km^2.

After a geothermal project positively exceeds the prefeasibility stage, it evolves to the feasibility stage. The verification of the existence of a reservoir in an area between 10 and 100 km^2 is carried out by drilling deep exploratory wells. Likewise, the probable reserve of the prospective area is evaluated, and the preliminary design of the systems to be used in the following stages is developed.

The development stage includes the continuation of well drilling, the execution of geoscientific studies in detail, the evaluation of the probable reserve, the extraction of the fluid, the design of the project, and the construction of a power plant generation.

The exploitation process of the resource involves the management of the geothermal fluid from its extraction to its exploitation through the production of electrical energy or any direct use of heat. Techniques are applied during the exploitation to optimize the use of the fluid and its transport to the generation plants, guaranteeing the continuous operation of the field and observing and controlling the reservoir evolution.

At the international level, there is currently no guide available that contains the most recommendable practices for the development of a geothermal project. Alone, with the experience acquired by the different countries, geothermal projects have been executed. The International Geothermal Association (IGA) [6] configured some ideas regarding the practices applied in different geothermal projects on the planet. Even though there are differences in methodologies and techniques between the different countries, the general guide establishes a seven-stage process for the development of these projects. This guide is known as ESMAP Geothermal Handbook and basically contains the stages mentioned by [5, 7], with the exception that in the exploitation stage, it adds two more stages: (1) construction of the generation plant and (2) start-up and operation. The seven-stage process proposed by the [6] is shown in **Figure 2**, adding at the end of seventh, the corresponding operation and field maintenance.

Both documents [5–7] coincide in the preliminary survey study as the basis for initiating a geothermal project. This document describes this stage in a specific way, considering the influence of the exploratory stage in a geothermal project.

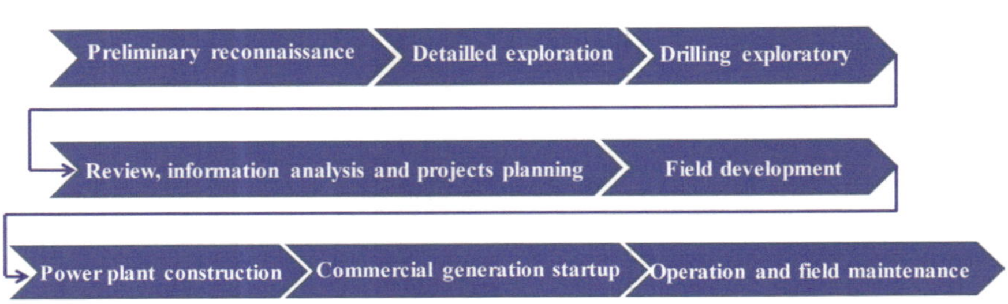

Figure 2.
Process stages of a geothermal project suggested by [5–7] and adding the operation and field maintenance stage.

2.1 Geothermal survey study

The preliminary reconnaissance stage involves a work program to evaluate the available evidence of the geothermal potential within a specific area. This stage is usually at the regional or even at the national level, where reviews of the literature on geology, hydrology, thermal manifestation data, well data, remote sensing data (if available), and even anecdotal information from the communities are included.

The studies that are carried out are the geological, geochemical, and hydrological exploration; using the available geophysical information also correlated the satellite images. A preliminary geological scheme of the geothermal system of the area or areas of interest is elaborated, and the geothermal possibilities of each are determined. The parameters to be determined are:

- The existence of an anomalous high-temperature zone in the subsoil based on the volcanological conditions and the result of chemical geothermometers.
- The degree of permeability of the rocks in which it is inferred the reservoir is nested. This determination will be based on the degree of tectonic fracturing suffered by the rocks.
- The extension and depth of the probable reservoir, which is used to estimate the volume of the deposit and is the sustenance to evaluate the energy capacity of the area, as well as the cost of the wells to be drilled.
- Estimate the possibility of recharging the aquifer that may sustain the reservoir exploitation during its operational life.

The analysis of the areas studied in this stage will allow to evaluate its energetic potential and a preliminary hierarchy of those that are considered of geothermal interest. Subsequent socioeconomic and political evaluations may help to define priority areas to continue prefeasibility studies.

In general, the geological studies are summarized:

- Develop regional cartography and define the preliminary geovolcanological scheme of the area.
- Define the relationship of regional geodynamics with tectonics and volcanism in the area.
- Determine thermal anomalies.
- Define the regional stratigraphic sequence and the lithological characteristics of the formations.
- Prepare the geovolcanological cartography of the geothermal areas identified.
- Identify the elements that could integrate the geothermal systems (heat source, reservoir, sealing system).
- Define, classify, and select geothermal areas of interest.
- Preparation of a preliminary structural geological plan of the region to be investigated.

From the specimens of collected rocks, thin sheets were prepared for petrographic and petrological analyses. The petrographic analysis is useful for determining the type of rock and classifying it. The petrological analysis allows to determine the nature of the magma and its degree of acidity.

Analysis of hydrothermal alteration will allow to determine the paragenesis of minerals and a thermal zoning of the possible geothermal system. The rock samples are subjected to chemical analysis for determining major elements such as SiO_2, AlO_3, FeO, Na_2O, CaO, MnO, TiO_2, P_2O_3, and MgO, among others, and also trace elements such as Ni, Cr, Rb, Sr., Ce, La, Zr, and Nb. By isotopy, the Sr^{16}/Sr^{37} ratio will be determined.

At this stage, the reservoir temperature could be estimated from the geochemical samples, using geothermometers or also taking into account the mineralogical studies of the hydrothermalized xenoliths, the petrology of the expelled products, and their radiometric age. The stratigraphic sequence of the studied zone will also contribute to define the hydrological characteristics of the formations.

Regardless of the studies to be carried out at this stage, it is highly recommended to visualize the project in a global manner, covering aspects such as infrastructure of the area (access roads, water availability, communication, and transmission systems) and property rights (water, legal permits for drilling, commitments to environmental legislation, legal framework of the geothermal laws, etc.). The consideration of these factors allows to identify possible obstacles that could influence the successful development of the project in general. Taking into consideration the results of the review and evaluation, the developer is in a position to decide to move to the prefeasibility stage. This stage could be developed in about 4 to 5 months; however, if there are many potential sites and if the processes of environmental approvals and permits are complex, it could last up to a year.

2.2 Prefeasibility studies

The basic studies of this stage are summarized in **Figure 3**.

The time duration for the recognition and prefeasibility stages, corresponding to the exploration stage [6] assigned in its document, is estimated to be 10 months. **Table 2** shows the breakdown of the activities of each stage using a Gantt chart for distribution over time.

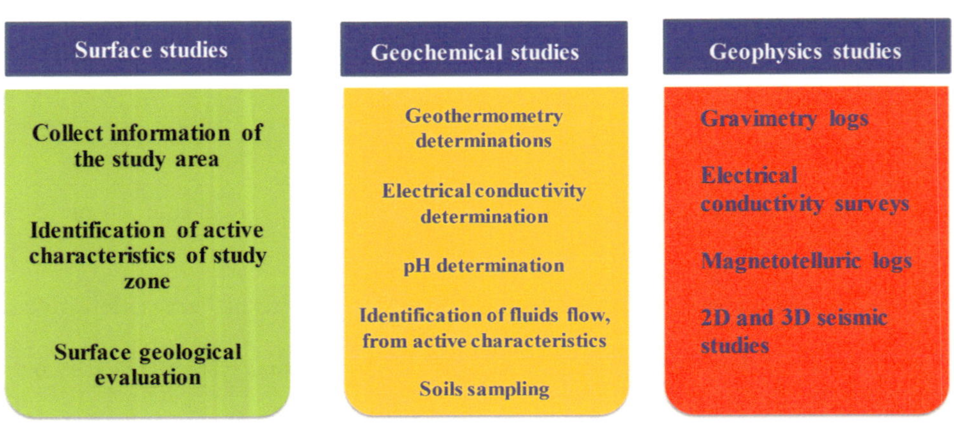

Figure 3.
Summary of the studies carried out through the three basic disciplines that intervene in this stage of superficial recognition of a geothermal project.

Stage	Activities	Time (months)									
		1	2	3	4	5	6	7	8	9	10
Preliminary reconnaissance	Review of updated literature	■									
	Access and land domain	■									
	Review and environmental and social problem identification	■	■								
	Attention to legal regulations	■	■								
	Inventory of active geothermal zones		■								
	Field general exploration program		■	■							
Exploration	Environmental study program according to regulation		■	■							
	Field works of geology and geochemical exploration		■	■							
	Geophysical surveys			■	■						
	Data interpretation by correlating geological, geophysical, and geochemical				■	■					
	Drilling of temperature gradient wells					■	■				
	Integrated analysis of information (geological, geophysical, geochemical, and thermodynamic)						■	■			
	Conceptual model development	■	■	■	■	■	■	■	■	■	■
	Site selection for drilling exploratory of deep well								■	■	
	Prefeasibility report							■	■	■	■

Table 2.
Sequence of the activities that are carried out in the stages of preliminary recognition and exploration, each with their duration times.

The main activities of each one of the disciplines that are executed throughout the exploration stage are the following:
Location of the source (lat/long or UTM coordinates).

1. For active geothermal characteristics

- Temperature
- Electric conductivity
- pH
- Flow
- Presence of gas bubbles and their composition
- Presence of odors (sulfides or others)
- Presence of precipitates in fluids
- Local or detailed area map with the thermal characteristics clearly indicated

2. Geological data

- Geological maps of the area or areas
- Cross-sectional geological sections of the study area or areas
- Description by means of stratigraphic and lithological columns
- Description of regional and local structures accompanied by maps
- Identification and characterization of the potential of the reservoir units
- Presence of mineralization associated with hydrothermal systems

3. Geochemical data

- Location, name, and characteristics of the sampling points
- Temperature, electrical conductivity, pH, and approximate flow during sampling
- Preservation of the sample for transfer to laboratories
- Chemical analysis of the samples collected
- Characteristics of the laboratories that develop the analyses
- Information on calcite inhibitors (if the sample comes from a producer well)
- Names, descriptions, and locations of the incrustations or minerals deposited

- Geothermometry estimates
- Interpretation and/or graphs of geochemical data
- Referral data from wells and neighboring projects

4. Geophysical data

 - Gravity logs
 - Electrical resistivity
 - Seismic surveys
 - Heat flow and temperature gradient records
 - Geomagnetic studies
 - Other records

5. Subsurface temperature data

 - Rock formation temperature from measurements
 - Temperature to flow conditions, at manifestations as in wells
 - Maps of temperature contours at different depths of the reservoir
 - Transverse sections of temperature distributions

6. Reservoir conceptual model that incorporates all the previous concepts.

All the data integration will allow the project developers to understand and explain the three-dimensional composition of the project area including its geological structure, stratigraphy, geophysical properties, location of the thermal sources, and geochemical characteristics. The model's fundamental result is the location of the heat sources and the possible sources of geothermal fluids as well as the nature of the trajectories that allow the movement of the fluids from the source to the system through the reservoir and to the discharge points.

Based on the understanding of fluid flow, the conceptual model is used for programing well drilling to reach specific targets on its lithology and structures. A robust conceptual model allows determining the feasibility of a project, the ideal locations, and depths to drill in the wells. The basic focuses of this type of models are oriented to:

- Determine temperature thermodynamic conditions for the generation of electrical energy.
- Determine the distribution of the temperature surfaces in the reservoir and its relationship with the fluid flow.

- Determine the adequate porosity and permeability existing in the rocks of the reservoir, as well as its extension and orientation.
- Determine the existing relationship between geology and geothermal characteristics.
- Determine the degree of influence of faults and stratigraphic units on the flow of fluids.
- Determine the most attractive places and their depths to succeed in drilling wells.

The final process of the exploration consists of the evaluation of all the data, technical and nontechnical, before committing the decision of a location to start the drilling of a deep well. This is an important decision because it involves large financial commitments for the project.

A geothermal project presents risks of success in its early stages, because geochemical studies, geological exploration, and geophysics are developed from the surface. Even though these studies are of very good precision, there is always the uncertainty of their validation. The previous is achieved with the drilling of a well, which represents a direct view of the formation of rock and its thermodynamic characteristics, confirming the veracity of the preceding studies. Conversely, the first stages do not involve large costs until the drilling begins. However, drilling influences the decrease in project risks. **Figure 4** shows a graph of the behavior of risks in a geothermal project [1, 8] and its relationship with investment costs according to the stages of the same. Ref. [9] developed an estimate of installed capacity costs for plants from 20 to 50 MWe in New Zealand.

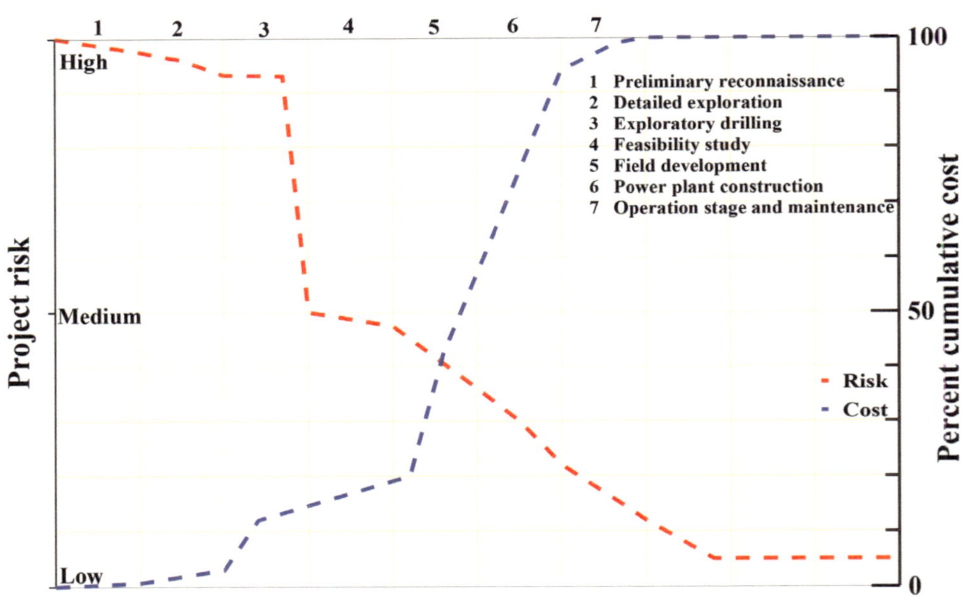

Figure 4.
Typical profiles of risks and costs in a geothermal project, taking into account each stage of its development [1, 8].

2.3 Well drilling

The drilling of the first wells in any project represents the period of greatest risk. It is usually recommended to drill three or four deep wells to obtain certainty of the commercial feasibility of the exploitation of the resource. Likewise, the drilling of at least one injection well must be considered jointly. It is important to consider that drilling wells, their logs, and their tests significantly improve the understanding of the available resource.

In this stage, they represent a reliable basis for preliminary estimation of the energy reserve, the identification of the production thicknesses, the petrophysical characteristics of the formation, and an approximate evaluation of the system area. The basic information that must be carefully collected at this stage is the following:

- Drilling speeds.
- The inlet and outlet temperatures of the drilling fluids.
- Chemical sampling of fluids and gases that come out with the circulation of the drilling fluid.
- Sampling (meter by meter) of the perforation cuttings for further analysis.
- Mineralogical control.
- Lithology control of the well (each meter), for the construction of its column.
- Record of circulation losses of drilling fluids.
- Series of records of temperature and pressure profiles approximately every 400 m, using resting times of 0, 6, 12, 18, 24, and 30 h to determine the profiles in undisturbed state and the static pressures and temperatures to be used in the numerical models of the reservoir.
- Injection tests when the target depth is reached in the well, to analyze them by means of pressure transient techniques and determine the petrophysical characteristics of the reservoir area.
- If the well presents thermodynamic characteristics of production, enable it with its surface facilities, valves, pipes, silencer, and measuring instruments (manometers, thermometers, and sampling holes) to make a discharge test to different openings and build the initial characteristic curve of the well production.
- During the production test, take samples of the fluid produced at each orifice change, for its chemical characterization.

The series of pressure-temperature logs, together with all the parameters that have been determined, allow to identify the characteristics of the well to make an appropriate definition on the presence of the appropriate range of exploitation. After the well drilling reaches the intervals of interest, transient pressure tests are carried out. The decision on well completion is taken after the permeable zones and thermodynamic conditions suitable for exploitation are found. During well completion, design of production pipes that are going to be introduced into the well is carried out. The next step is to proceed to conditioning the facilities on the surface of the well, giving a period of heating for their pipes and even measuring their elongation on the surface.

Performing daily measurements on the pressure and temperature at the well head, it is possible to determine if the heating has been uniformed in the pipes, in order to decide the appropriate moment to carry out the initial discharge test and obtain its production characteristic curve. The parameters of the characteristic curve (pressure-flow-enthalpy) are the basis for the well characterization.

Taking all previous considerations, the results of the first wells are aimed at:

- Making a better approach to the heat source
- Determining the average productivity as well as the probability that the same trend will continue in future drilling
- Selection of new sites, target depths, and construction of road infrastructure
- Mechanical designs of the subsequent exploitation and reinjection wells
- Development of a preliminary design of the power plant and the steam transport system

2.4 Project planning

The first drilled wells in an area allow locating and confirming the existence of a viable project, and from the information recovered, their risks are substantially reduced. With all the information obtained, it is feasible to prepare a report. At this point the developer is able to figure out the general plan of the project. The incorporation of the risk analysis in the economic models of the project provides support for the request of the financing that was necessary.

The feasibility report is designed in order to provide confidence on the viability of the project and facilitate its financing. From this stage and in the subsequent ones, a constant monitoring of the behavior parameters of the wells must be maintained, such as:

- Production behavior (pressure, flow, enthalpy, production hole diameter, etc.)
- Fluid and gas sampling (chemical and isotopic analyses)
- Production tests (according to the possibilities in each well)

The feasibility report should include the following elements:

- Location of wells to be drilled, design for completion, and civil works necessary (access roads, leveling, possible locations of the generation plant, etc.)
- Design of field development wells
- Design of the electric power generation plant

- Construction of the main access roads
- Budgeting
- Terms of agreements for the sale of electric power
- Projections of the income-expenditure budget

2.5 Field development

Field development corresponds to its growth through production wells for increasing production flow until the supply requirement to the generation plant is satisfied. Also the environment-friendly eviction must be sought, of the brine produced, through reinjection wells.

The data and useful information that project partners must analyze may include the following:

2.5.1 Information related to drilling equipment, work crews, and supervisors

- Detail specifications of the drilling equipment for the subsequent wells
- Analysis of the experience, in geothermal wells, of the drilling company
- Design of drilling programs for the subsequent wells to be drilled (depths, thicknesses, casing pipes, production liners, surface installations, etc.)
- List and classification of possible specialized service providers
- Knowledge and classification of the work teams for the project and the technical support of their advisors

2.5.2 Information on the commercial aspects of the project

- Plan for the sale of electric power and its sale wholesale or retail
- Description of the project structure
- Analysis of financial projections including expenses for unforeseen contingencies during drilling or costs of maintenance operations and/or stimulation
- Description and value of the goods and equipment to be insured

2.5.3 Information of the company responsible for the project

- Basic information of the company (type, organization, owners, financial movements, related companies, etc.)
- Experience of managers in the management of drilling projects and the development of natural resource projects

	Exploration			Exploitation		
Phases	Year 1	Year 2	Year 3	Year 4	Year 5	Year 6
	Recognition	Exploration	Exploratory drilling	Feasibility	Field development	Construction
Generic activities	Licenses acquisition	Geology detailed	Drilling and evaluation of exploration wells	Financing analysis	Well drilling	Construction of the plant
	Regional surface geological exploration	Geochemistry	Parameter evaluation	Conceptual design	Exploitation designs	Drilling and evaluation of wells
	Geochemical exploration	Geophysical studies	Static reservoir characterization	Reservoir evaluation	Plant design	Construction of vapor ducts and surface equipment
		Location of sites to drill	Delimitation of the exploration area	Feasibility report	Design of steam transport and equipment	Admission tests and turbine treading
Main objectives	Surface recognition	Confirmation of geothermal conditions	Geothermal reservoir confirmation	Techno-economic feasibility	Techno-economic infrastructure	Commercial operation start-up

Table 3.
Activities and their likely duration of the various stages of a business plan for a geothermal project [1, 5].

- Experience of the work team
- Number and type of support staff

2.6 Construction of the power generation plant

Based on the construction designs of the generation plant, at the end of this stage, civil works, access roads, and steam collection and transport systems are completed. During this stage, the development of the transient pressure tests to the wells that are finishing drilling is continued as far as possible. These tests are useful for reservoir characterization under its undisturbed state.

2.7 Start-up and operation

After finishing the construction of the plant, the start-up and operation phase of the plant begins. Having assured the supply of the resource (with the drilling of wells) for the start of operation of the plant, the main objective is aimed at optimizing production. The above is achieved through appropriate designs, applying criteria for reservoir engineering and well production.

Likewise, it is necessary to emphasize the drilling of injection wells for the clean evacuation of the brine, preserving the environment and seeking reservoir recharge. During the operation, it must seek to minimize its costs and ensure reliable delivery of the resource to the generation plant.

A timeline on the likely total duration of the project and each of its stages is shown in **Table 3** [1, 5]. It should be emphasized that the quality of the information during the first two stages is the basis for obtaining a location for the first well to be drilled and, if necessary, ensuring its success and therefore the continuity of the project. However, in the case that the first well drilled is not a producer, it is recommended not to discard the project and try to drill up to three or four exploratory wells, always applying the best technical criteria to select their locations.

After the exploitation stage comes the operation and maintenance stage. This stage involves the operation and maintenance activities of the electricity generation plant, as well as the maintenance of the wells, the transport systems, and the accessory equipment for the operation of the plant.

3. Costs

According to the Department of Information and Energy Administration of the United States of America (EIA, 2016), the costs that generally prevail at present for a geothermal plant are the following:

The estimated cost of the KW installed in a geothermal plant fluctuates between 1500 and 2500 dollars. The cost of operation and maintenance of a plant can be considered between 0.015 and 0.020 dollars per KW installed hour. The cost per KW hour installed for the operation and maintenance of the field is estimated between 0.035 and 0.040 dollars. However, the magnitude of generation plant type is a factor for investment projection, because the initial drilling exploration increases the costs. **Table 4** shows the investment costs for power plants of different generation capacities. From cost balance, it can be seen that the cost of generated MW for small plants (5 MW) is major than for bigger plants (30 MW).

Plant capacity	Stage	High-enthalpy fluid	Medium-enthalpy fluid
Plants with capacity lower than 5 MW	Exploration (USD)	400–800	400–1000
	Production (USD)	100–200	300–600
	Operation (USD)	1100–1300	1100–1400
	Total (USD)	**1600–2300**	**1800–3000**
Plants of medium capacity between 5 to 30 MW	Exploration (USD)	250–400	250–600
	Production (USD)	200–500	400–700
	Operation (USD)	850–1200	950–1200
	Total (USD)	**1300–2100**	**1600–2500**
Plants with capacity upper than 30 MW	Exploration (USD)	100–200	100–400
	Production (USD)	300–450	400–700
	Operation (USD)	750–1100	850–1100
	Total (USD)	**1150–1750**	**1350–2200**

Table 4.
Estimated values of investment of a complete geothermal project from exploration to operation and maintenance stage, with generation plants of 5 up to 30 MW and for fluids with high and medium enthalpy [10].

The useful life of a geothermal project is estimated between 25 and 30 years. The considerations were made based on a 50 MW plant generating electricity using geothermal fluid.

4. Direct uses

The direct uses of geothermal energy are considered among the oldest, most versatile, and most common ways to use it. Since the Roman Empire, there are records of using water for thermal spas [11–13]. Other applications were for space heating, and then techniques were introduced to use it as a greenhouse effect; recently it has also been applied for refrigeration, for fruit drying, and with the advancement of technology in heat pumps.

Table 5 shows the use of geothermal energy applied to direct uses in the last 20 years on the planet [13]. It can be seen that the greatest growth has occurred due to the technological development.

In addition to this table, a growth in the application of geothermal energy for heat pumps, heating and greenhouses, thermal baths, and swimming can be observed (**Table 5**). However, the drying of fruits, vegetables, and woods, as well as applications for refrigeration and industrial uses, shows moderate progress.

The geothermal fluid used for direct applications has low enthalpy, and the decision to operate under this modality is based on two of the basic stages of the project:

1. At the end of the exploration stage, when the prefeasibility study indicates the geothermal project is not viable so that the wells can be subjected to commercial exploitation.

2. During the stage of commercial exploitation, the use of the geothermal resource makes the process interesting to take advantage of the remaining energy in each stage.

Direct uses of geothermal heat	Utilization TJ/year				
	1995	2000	2005	2010	2015
Heat geothermal pumps	14,617	23,275	87,503	200,149	325,028
Heating of spaces	38,230	42,926	55,256	63,025	88,222
Air conditioning of greenhouses	15,742	17,864	20,661	23,264	26,662
Heating of aquaculture ponds	13,493	11,733	10,976	11,521	11,958
Drying of fruits, vegetables, and woods	1124	1038	2013	1635	2030
Industrial uses	10,120	10,220	10,868	11,745	10,453
Thermal baths and swimming	15,742	79,546	83,018	109,410	119,381
Refrigeration	1124	1063	2032	2126	2600
Others	2249	3034	1045	955	1452

Table 5.
Description of the uses of geothermal energy for direct applications on the planet during 1995–2015 [13].

That is, if you have an electric generation plant with an intake pressure of, for example, 8 bars, when you leave this first process, the steam still has pressure, and naturally there is water condensed at high temperature, resulting from the first "flash." The steam can be channeled to a second "flash" to a plant that requires lower intake pressure, and again there is condensed water at high temperature. After this second stage, the steam still has energy and can be channeled to a binary cycle plant to take advantage of its remaining energy. The water produced by the consecutive "flashes," in the generation processes at different intake pressures, can be used to heat spaces in residential areas. When the water cools, it can be conducted to the injection systems to recharge the reservoir.

As can be seen, the direct uses of geothermal energy have a variety of applications during the operation of the field, which leads to make the most of the heat of the earth while preserving the environment. For this reason, the geothermal energy is renewable and can be considered sustainable.

5. Conclusions

Geothermal energy is a natural resource whose characteristics make it renewable and sustainable.

The exploration, geological, geochemical, and geophysical stages allow us to delimit the regional area to a focused area where we can locate the first exploratory wells to be drilled.

The quality of the information obtained, its analysis, and interpretation support the technical bases of the stages of recognition, prefeasibility, and feasibility of a geothermal project.

The risks in the exploration stage are high due to the uncertainties of not yet having direct information from the subsurface; however, the costs are low, which may be 15 or 20% of the total cost of the project.

The costs of the project increase during the well drilling stage for the development of the field, as well as in the construction of the plant; however, the risks decrease noticeably.

The characterization of the productivity of the wells allows to establish the appropriate designs of their exploitation and in this way increase their efficiency and useful life.

As of the start-up of the plant, costs are minimized, because the working fuel is a natural resource; only operating and maintenance costs are considered for wells and facilities. It is necessary to maintain a balance between the extracted mass and the recharge input to increase the efficiency in the operation.

In a geothermal process, particularly a hydrothermal of high temperature, the energy can be fully utilized by sequentially using its discharges from the different stages, to generate electricity at lower acceptance conditions of the plants in a second and third "flash."

Direct uses of geothermal heat are recommended for low-enthalpy systems.

Acknowledgements

The authors express their gratitude to the National Institute of Electricity and Clean Energies of SENER, México, for the support.

Author details

Alfonso Aragón-Aguilar[1]*, Georgina Izquierdo-Montalvo[1], Daniel Octavio Aragón-Gaspar[2] and Denise N. Barreto-Rivera[3]

1 Instituto Nacional de Electricidad y Energías Limpias, Cuernavaca, Morelos, México

2 Facultad de Arquitectura, Universidad Autónoma del Estado de Morelos, Cuernavaca, Morelos, México

3 Instituto de Ciencias de la Educación, Universidad Autónoma del Estado de Morelos, Cuernavaca, Morelos, México

*Address all correspondence to: aaragon@ineel.mx

References

[1] ESMAP (Energy Sector Management Assistance Program). Manual de geotermia: Como planificar y financiar la generación de electricidad, Informe Técnico 002/12. Washington D.C., U.S.A: Grupo del Banco Mundial; 2012. 164 pp

[2] Bertani R. World geothermal power generation in the world 2005-2010 update report. Geothermics. 2012;**41**:29

[3] Bertani R. Geothermal Power Generation in the World 2010-2014 Update Report. In: Proceedings World Geothermal Congress, Melbourne, Australia; 2015. CD. 19 pp

[4] Richter A. Top 10 Geothermal Countries Basec on Installed Capacity—Year end 2017. ThinkGeoEnergy, January. Available from: http://www.thinkgeoenergy.com/top-10-geothermal-countries-based-on-installed-capacity [Accessed: May 16, 2018]

[5] OLADE (Organización LatinoAmericana de Energia), BID (Banco InterAmericano de Desarrollo). Guía para estudios de reconocimiento y prefactibilidad geotérmicos, Quito, Ecuador; 1994. 146 p

[6] IGA Service GmbH. Geothermal exploration best practices: A Guide to Resource Data Collection, Analysis, and Presentation for Geothermal Projects. Germany: Bochum University of Applied Sciences (Hochschule Bochum); 2013. 74 pp

[7] OLADE (Organización LatinoAmericana de Energia), BID (Banco InterAmericano de Desarrollo). Metodología OLADE para la explotación geotérmica, Serie de Documentos de OLADE, Quito, Ecuador; 1980

[8] IPPC. Special report on the intergovernmental panel on climate change. In: Managing the Risks of the Extreme Events and Disasters to Advance Climate Change Adaptation. Nueva York, EUA: Cambridge University Press; 2012. 582 pp

[9] Barnett P, Quinlivan P. Assesment of current costs of geothermal power generation in New Zealand, SKM Report for New Zealand Geothermal Association; 2009

[10] Jaimovich O. Costos de la energía geotérmica, Boletin 02/07, Geotermia. Universidad de Buenos Aires; 2007. 8 pp

[11] Mwangi MN. Planning of geothermal projects: A case study on Kenya. In: Short Course II on Surface Exploration for Geothermal Resources, UNU-GTP and KenGen, Lake Naivasha, Kenya; 2-17 November 2007

[12] Lund WJ, Freeston HD. Worldwide direct uses of geothermal energy 2000. In: Proceedings World Geothermal Congress, Kyushu-Tohoku, Japón, May–June; 2000. CD. 21 pp

[13] Lund WJ, Boyd TL. Direct utilization of geothermal energy 2015 worldwide review. In: Proceedings World Geothermal Congress, Melbourne Australia; April, 2015. CD. 31 pp

Chapter 5

Space Cooling by Ground Source Heat Pump in Tropical Asia

Kasumi Yasukawa and Youhei Uchida

Abstract

In Southeastern Asia, where energy demand is expanding to meet the increasing population and industry needs, energy saving by use of ground source heat pump (GSHP) could be one of the solutions. There are several concerns on GSHP installation in this region. The biggest concern is the subsurface temperature in tropical Asia. Although space cooling is needed in tropical regions, underground is slightly warmer than average atmospheric temperature and may not be used as "cold" source. However, groundwater temperature survey results in Thailand and Vietnam show the applicability of GSHPs in this region. Also, experimental GSHP systems for cooling have been installed in Thailand, Indonesia, and Vietnam, and studies have been done to improve cost performance of these systems. As results, the following things are found: 30% of energy saving compared to normal air-conditioner has been confirmed at a test site in Bangkok. Systems with local manufacturing would be a key for cost reduction. Cost performance may be optimized by selection of horizontal and/or vertical heat exchangers depending on the local subsurface condition. Drilling technology for no-cementing and no-casing completion is a key for higher heat exchange rate in vertical heat exchangers.

Keywords: ground source heat pump (GSHP), tropical Asia, space cooling, groundwater flow, temperature survey, advection effect, apparent thermal conductivity, drilling, no cementing, polymer

1. Introduction

Ground source heat pump (GSHP) system for heating and cooling purposes may be a powerful alternative to reduce energy consumption and to contribute to environmental issues. Its intensive utilization for heating may reduce emissions of CO_2 and other toxic gases by replacing fossil fuel boilers.

It may also greatly contribute to mitigate the urban heat island (UHI) phenomenon, since GSHP operation for cooling does not emit waste heat to atmosphere. UHI is a matter of great concern in megacities [1] because it triggers bad circulation of energy consumption. The higher the atmospheric temperature, the more energy is consumed for space cooling, resulting in even higher atmospheric temperature in urban areas (**Figure 1**). Expansive population growth and urbanization in Asia would make the problem of UHI more serious. In such a situation, intensive use of GSHP may largely contribute to cut the bad circulation of UHI. An estimation [2] shows that full installation of GSHP in the central part of Tokyo may reduce the daily maximum atmospheric temperature in the summer by 1.2 K through combined effects of high efficiency of GSHP and reduction of UHI.

However, in Southeastern Asian countries, where significant economic growth is expected so that energy saving and environmental protection will be major matters of importance for sustainable growth, the current number of GSHP installations is quite limited and rapid growth of GSHP installation is desirable.

GSHP has been considered not appropriate in tropical regions where only space cooling is needed. Since seasonal change of atmospheric temperature is quite limited in tropics and subsurface temperature is generally higher than year-average atmospheric temperature, underground may not be appropriate as a "cold heat-source" (**Figure 2**). Nevertheless, there still exist advantages of GSHP use in tropical regions if (1) daily changes of atmospheric temperature exist and (2) subsurface temperature is rather low and/or the advection effect of groundwater flow in shallow aquifer raises heat exchange rate.

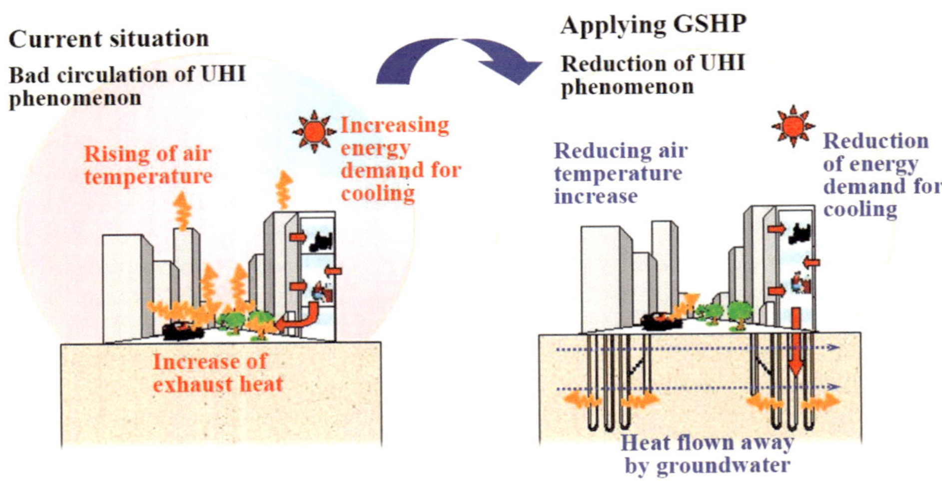

Figure 1.
Reduction of UHI phenomenon by intensive installation of GSHP systems.

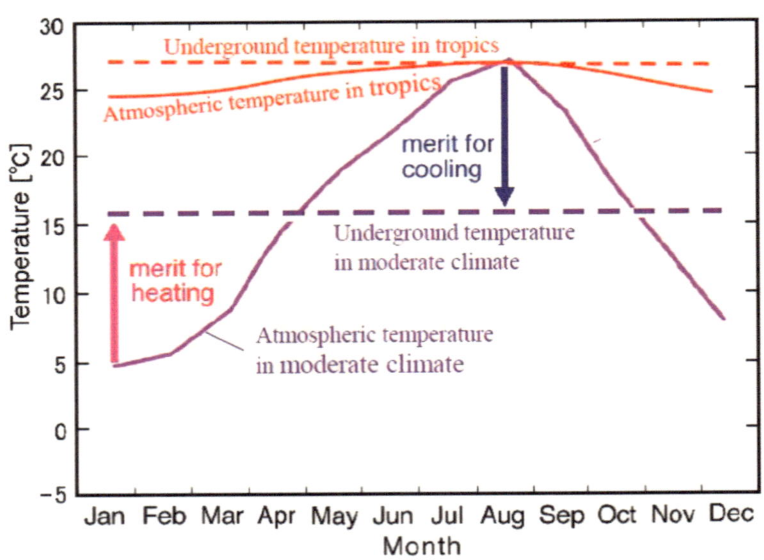

Figure 2.
Comparison of monthly average atmospheric temperature and underground temperature in moderate climate and tropical regions (conceptual figure).

According to a result of groundwater temperature survey conducted in the Chao Phraya plain, Thailand, subsurface temperature is lower than daytime atmospheric temperature throughout the year in most cities [3]. Considering that the major consumers of cooling systems in these areas are offices and shops, higher performance only in daytime would still have advantage to normal air-conditioners. It suggests that underground may be used as a cold heat-source even in parts of tropical regions.

Thus, to verify the applicability of GSHP in tropical regions, mapping of subsurface temperature is essential. Then, demonstration of the GSHP system is important to understand a possible performance of GSHP systems in the region and to make practical guidelines on system design and installation procedure. Therefore, in the following sections, subsurface temperature survey and results of experimental installations of the GSHP system in Thailand and in Vietnam are presented.

2. Subsurface temperature survey in tropical regions

2.1 Effects of natural subsurface temperature and groundwater flow on GSHP systems

For a GSHP system design, information on groundwater is quite important. For open-loop systems, in which groundwater is extracted for heat exchange at ground surface and re-injected afterward, information of aquifer on depth, temperature, and flow rate is important. For closed-loop systems in which heat exchange is conducted by circulating fluid in a heat exchange tube buried underground, information on subsurface temperature and apparent thermal conductivity is important (see **Table 1**). In monsoon Asia, where shallow groundwater flow is dominant, apparent thermal conductivity is largely affected by advection effects of groundwater flow. Therefore, information on subsurface temperature and groundwater flow is essential for both closed and open systems. Thus, in this subsection, effects of groundwater flow on subsurface temperature distribution and its implication on GSHP systems will be explained.

Subsurface temperature in natural state at a depth of 20 m or deeper is normally stable throughout the year and generally slightly higher than year-average atmospheric temperature of the place. **Figure** 2 schematically shows the seasonal variation of atmospheric and subsurface temperature at a depth of about 50 m. In moderate climate regions, where subsurface temperature is higher than atmospheric temperature in the winter and lower in the summer, the GSHP system is useful for both space heating and cooling. On the other hand, in tropics where

Term	Measurement method	Affecting matters	Notes
Thermal conductivity σ	Thermal conductivity of a dried rock sample	Rock property	Unconsolidated sediments have lower σ than hard rocks
Effective thermal conductivity σ_e	Thermal conductivity of a water-saturated rock sample	Rock and water properties	$\sigma < \sigma_e$
Apparent thermal conductivity σ_a	Thermal conductivity measured at a site	Rock and water properties and flow rate	$\sigma < \sigma_e < \sigma_a$ for saturated zone, higher for higher flow rate

Table 1.
Terms of thermal conductivity in this chapter.

space cooling is needed, subsurface temperature is higher or approximately equal to atmospheric one and no advantage of GSHP systems can be seen.

Nevertheless, there exist temperature variations at the same depth of a plain or a basin. Natural groundwater flow, controlled by the topography of the ground surface and subsurface boundaries of rock permeability, may perturb the subsurface thermal regime. At recharge zones, infiltration of precipitation disturbs heat conduction from a depth that lowers the shallow subsurface temperature, while upward groundwater flow encourages heat transfer from a depth at discharge zones. Therefore, within an identical groundwater system, subsurface temperature at recharge zone is generally lower than that at discharge zone as shown in **Figure 3**. Thus, temperature difference of a few Kelvins may be achieved by groundwater flow. Another aspect of groundwater flow is that recharge zones have higher flowing velocity because of their non-flat topography. In the central flat region of a plane, groundwater flows more slowly.

Groundwater flow has another important effect on subsurface heat exchange using a closed-loop GSHP system. Advection effect of groundwater flow reduces temperature rise/drop around the borehole during heat exchange, which otherwise degrades the system performance. Thus, groundwater flow contributes to sustainable operation of GSHP systems. Therefore, if subsurface layers are effectively cooled by groundwater flow, GSHP systems may be useful in tropical regions.

2.2 Temperature survey at the Chao Phraya plain, Thailand

The authors conducted groundwater temperature measurements in the Chao Phraya plain in numerous observation wells belonging to the Department of Groundwater Resources (DGR), Thailand, from 2003 to 2005 [3]. Locations of the observation wells are shown as red dots in **Figure 4**. It should be noted that temperature profile should be taken in observation wells, in which water temperature reaches equilibrium with subsurface temperature. Topographically, the Chao Phraya plain consists of upper plain and lower plain, and the groundwater system is separated into two flow systems as well at a border around N15°40'. Nakhon Sawan is a discharge zone of the upper plain where groundwater discharges into a lake and flows away as river water while Chai Nat is a recharge zone of the lower plain.

Figure 4 shows the observed temperature profiles in these wells for each region. Wells in the same region have similar temperature profiles. Temperature inversions, in which shallow subsurface temperature is lower than surface temperature, are commonly seen in profiles in Chai Nat and Chachoengsao regions probably because

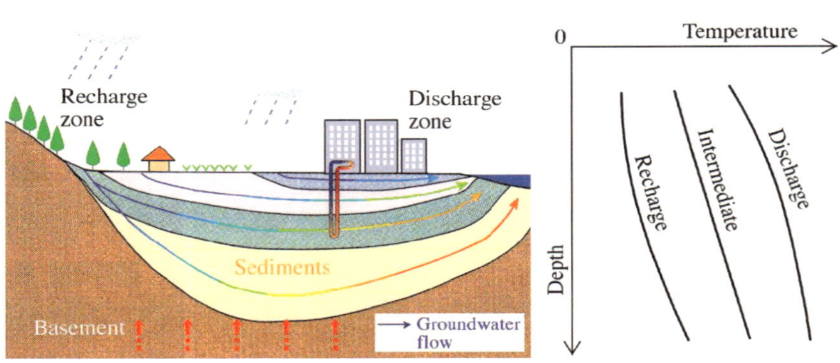

Figure 3.
Schematic image of groundwater flow in a plain and vertical temperature profiles in hydrologically different zones.

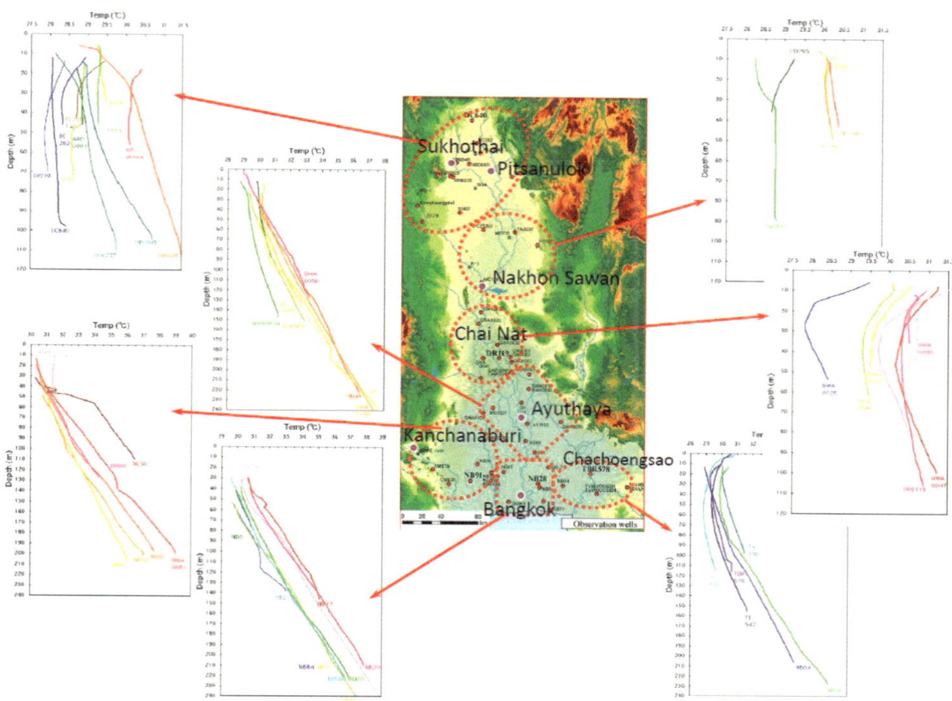

Figure 4.
Temperature profiles widely measured at observation wells in the Chao Phraya basin [3]. Red dots in the map show the locations of observation wells.

of global warming. Since surface temperature has risen in recent decades, it becomes higher than subsurface temperature that remains at the past level. This temperature inversion caused by global warming is typically observed in recharge zones [4, 5].

The proper depth of borehole heat exchanger for space cooling may be around 50 m or less, because subsurface temperature increases with depth and deeper parts are not appropriate as "cold" heat-source. For this reason, temperature range at depths between 20 and 50 m in each area is indicated in **Figure 5**. Temperature at a depth of 20 m or shallower is ignored because it may be affected by daily and seasonal changes so that observed values may not represent the statistical mean.

Figure 5 compares atmospheric and subsurface temperature at Bangkok, Ayutthaya, Nakhon Sawan, Phitsanulok, Sukhothai, and Kanchanaburi regions, respectively. For Phitsanulok and Sukhothai, an identical set of subsurface temperature data was used, while the atmospheric ones are different. At four regions out of six, Bangkok, Ayutthaya, Phitsanulok, and Nakhon Sawan, subsurface temperature is lower than monthly mean maximum atmospheric temperature throughout the year. The GSHP system may be effective in these areas for space cooling especially in daytime.

2.3 Temperature survey at the Red River plain, Vietnam

Groundwater temperature survey in the Red River plain, Vietnam, in observation wells belonging to the Department of Geology and Minerals of Vietnam (DGMV) was conducted in 2005 and 2006 [3]. Note that this region is not a tropical but semi-tropical area, but the knowledge obtained from this survey would be applied to tropical parts of the nation.

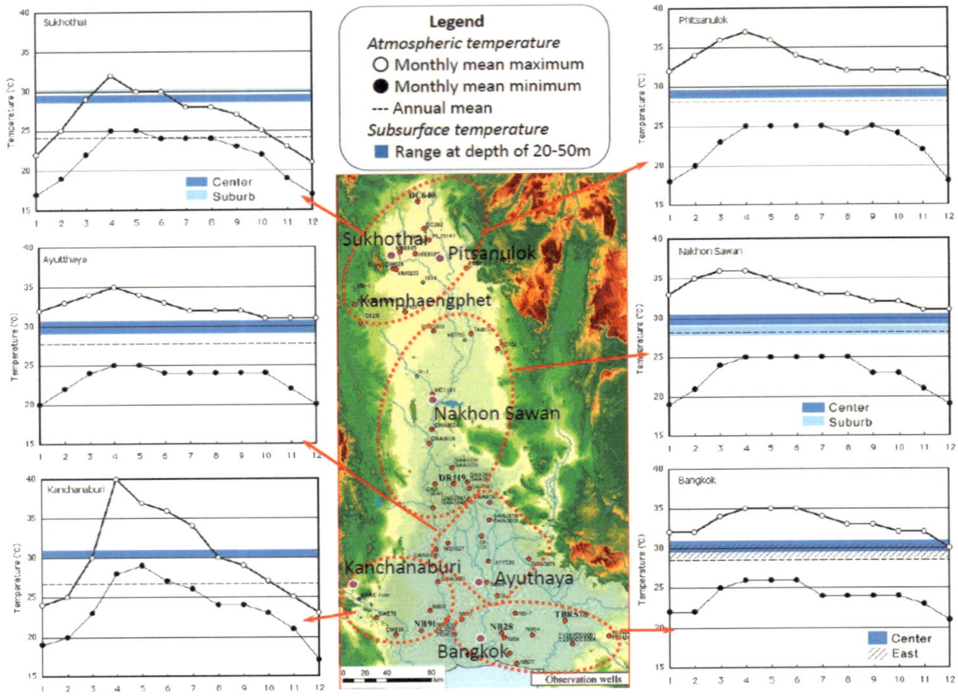

Figure 5.
Comparison of atmospheric and subsurface temperature at depths between 20 and 50 m at each region [3].

Location of observation wells in this area and their temperature profiles are shown in **Figure 6**. The color of each profile in **Figure 6** (right) corresponds to that of wells in **Figure 6** (left).

In the south of the Red River, the wells near the sea (Q108, Q109, and Q110) show higher temperature and temperature gradient than those in Hanoi, suggesting that the coast is a discharge zone while Hanoi is an intermediate zone of a groundwater system. The wells in the north of the Red River (Q131, Q159, Q158, and Q156) show lower temperature than those in Hanoi although they are nearer to the sea. But still, their temperature decreases with the distance from the sea. The groundwater system in the north must have a different origin from that in the south.

Figure 7 shows the monthly change of atmospheric temperature and subsurface temperature in Hanoi area. The subsurface temperature range is obtained from wells shown by circle in **Figure 6**, for depths of 20–50 m. In Hanoi, subsurface temperature is lower than monthly mean maximum atmospheric temperature from May to October by 2–7 K. Therefore, the underground may be used as a "cold heat-source" in the summer season.

2.4 Discussion on possibility of GSHP application

Based on the temperature observation results in Thailand and in Vietnam, the possibility of GSHP application for space cooling was identified for most places where subsurface temperature becomes lower than atmospheric temperature in daytime.

Generally, in a same source temperature condition, a water-source heat pump such as GSHP has higher performance than an air-source heat pump such as conventional air-conditioner because of its high heat exchange rate. A literature review [6] shows that their open GSHP gives higher coefficient of performance

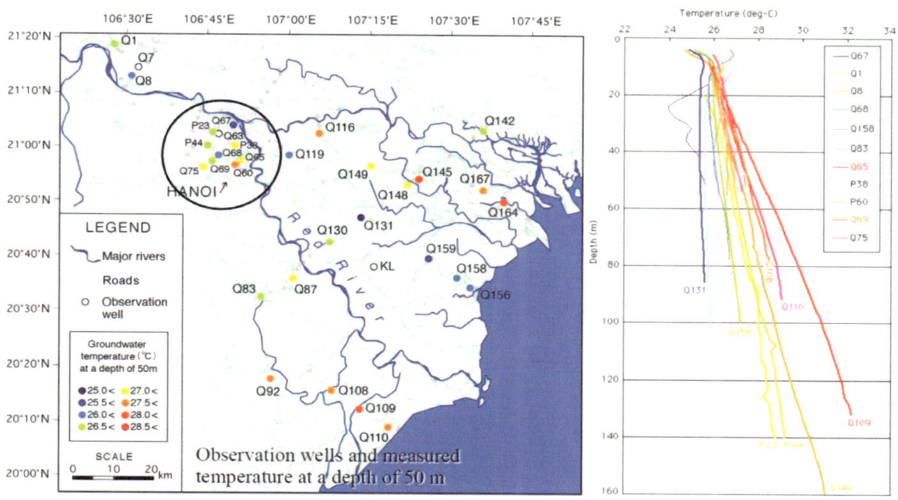

Figure 6.
Location of observation wells around the Red River, Vietnam, (left) and their temperature profiles (right) [3].

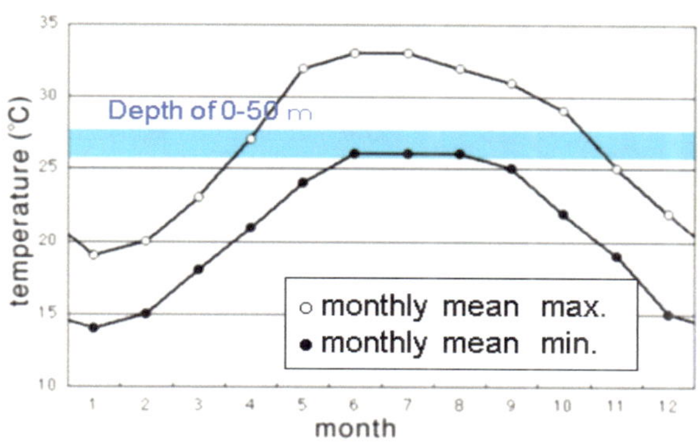

Figure 7.
Comparison of atmospheric and subsurface temperature at depths between 20 and 50 m at Hanoi area [3].

[COP, (generated heat)/(electricity consumption of heat pump)] than an air-source heat pump for cooling even when the atmospheric temperature is lower than ground temperature by 3 K. They suggest that, in their case, the COP of GSHP and air-source heat pump may be equivalent when the atmospheric temperature is lower than ground temperature by 5 K. Applying this result to **Figure 5**, a GSHP may have higher COP than an air-source heat pump in most places for most seasons in day time.

3. Demonstrations in Thailand and in Vietnam

3.1 Introduction

The authors installed and operated several experimental and demonstrational GSHP systems in tropical Asia. All these systems are closed-loop systems aiming at easier installation at lower cost. **Table 2** shows the essence of these experiments and demonstrations. The earlier systems are only experimental and removed after a year or more of operation. The later ones are demonstrational systems that have been continuously used by the people working at the site. Improvements have been done through these installations for better cost performance, which include drilling and well completion technology to achieve higher heat exchange rate, a combination of horizontal and vertical heat exchangers, and a controlling system for the heat pump operation. Details of the first experiment will be described in the next subsection and improvements in the other installations will be explained in the following subsection.

No., Place	Period	Subsurface heat exchanger	Surface system	Note
1. Kamphaengphet (DGR), Thailand	October 2006 to March 2008	57-m deep borehole with double U-tube	Water-water chiller, fan coil	First experiment in tropics. Mostly made in Japan.
2. Chiang Mai (DGR), Thailand	March 2008 to July 2010	80-m deep borehole with single U-tube + 60-m horizontal tube	Same as above	Moving the above system to another site.
3. Bangkok (Kasetsart Univ.), Thailand	July 2010 to 2012	200-m horizontal tube	Same as above	Moving the above system to another site.
4. Bandung (ITB), Indonesia	July 2013 to 2015	200-m horizontal tube	Remodel from air-conditioner	Cooling efficiency 25% up. Done by Akita University.
5. Bangkok (Chulalongkorn Univ.), Thailand	May 2014 to present	50-m deep borehole with single U-tube × 3 (150 m)	Combined chiller and fan unit	Cooling efficiency 30% higher than normal air-conditioner.
6. Bandung (Western Java Energy Mineral Institute), Indonesia	(pending: planned in 2015)	100-m deep borehole with single U-tube (installed in 2017)	Remodel from air-conditioner (planned)	Heat pump made in Indonesia is expected.
7. Saraburi (Chulalongkorn Univ.), Thailand	June 2015 to present November 2016 to present	300-m carpet style (horizontal) 300-m coil style (horizontal)	Combined chiller and fan unit	Machine made in Thailand Remodel from air-conditioner.
8. Pathumthani (Geology Museum, DMR) Thailand	March 2015 to present	50-m deep borehole with double U-tube × 2 (400 m)	Combined chiller and fan unit	Mostly made in Japan. No cementing borehole for higher heat exchange.
9. Hanoi (VIGMR), Vietnam	October 2016 to present	50-m deep borehole with double U-tube × 2 (400 m)	Combined chiller and fan unit	Mostly made in Japan. No cementing borehole for higher heat exchange.

Table 2.
Experiments and demonstration of GSHP cooling in tropical region by AIST and/or Akita University in collaboration with local institute.

3.2 Kamphaengphet experiment

An experimental GSHP system for space cooling was installed in DGR Kamphaengphet office, Thailand, in 2006 and operated for 17 months. Kamphaengphet is located at the edge of the Chao Phraya basin where groundwater flow is rather high (**Figure 5**). **Figure 8** (left) shows the temperature profile of an observation well in Kamphaengphet office measured before the installation of the GSHP system. The temperature range below water level is 30.1–30.6°C. A comparison of atmospheric and subsurface temperature at this place is shown in **Figure 8** (right). Except for December, subsurface temperature is lower than monthly mean maximum atmospheric temperature.

Figures 9 and **10** show the installed system. A heat exchange borehole (well) was drilled to a depth of 56 m and completed by normal cementing and a screen at the bottom. Then, a double U-tube (heat exchange pipe) was inserted and the borehole was filled with pebbles so that the water level in the borehole may keep equilibrium with the water-head of the bottom hole. The circulation fluid of both primary and secondary loops is just water because no brine is needed in tropical regions.

All major materials such as heat pump, fan-coil, and U-tube were exported from Japan although they are costly. Thus, the purpose of this project is to confirm genuine technical feasibility of GSHP in tropical regions, and the feasibility analysis of local material and cost was not included in this project scheme. Surface piping was made by normal PVC pipes which were purchased in local shops.

Temperatures in the subsurface heat exchanger, the room for cooling, and atmosphere were monitored in this project. The electricity consumption of the whole system and temperatures and flow rates of primary and secondary fluids were measured as well to calculate the system coefficient of performance (SCOP). For more details of this experiment, literatures are available [7, 8].

The system was operated in working hours of the office, approximately 8 hours a day in daytime, 5 days a week. Several times, the system operation was stopped for a week or longer for maintenance and the temperature data during these periods were effectively used to evaluate short-term and long-term effects of this system operation on subsurface temperature.

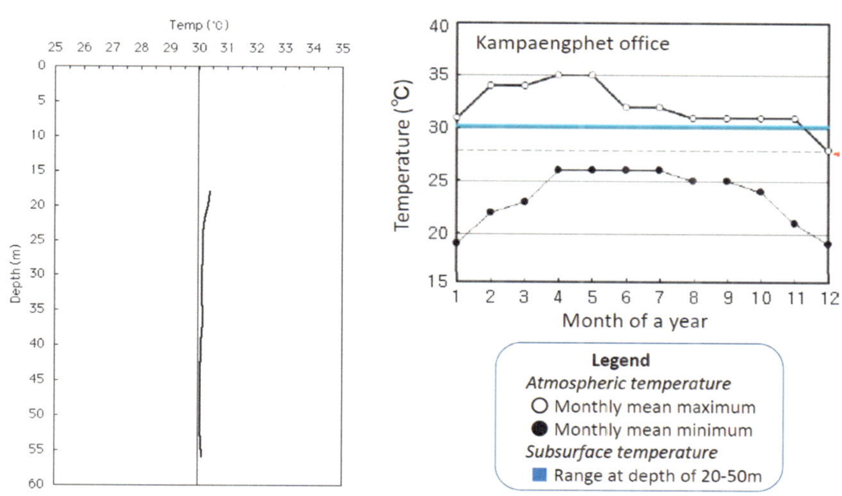

Figure 8.
Temperature profile of an observation well in Kamphaengphet office measured in December 2005 (left) and comparison of atmospheric and subsurface temperature (right) [7].

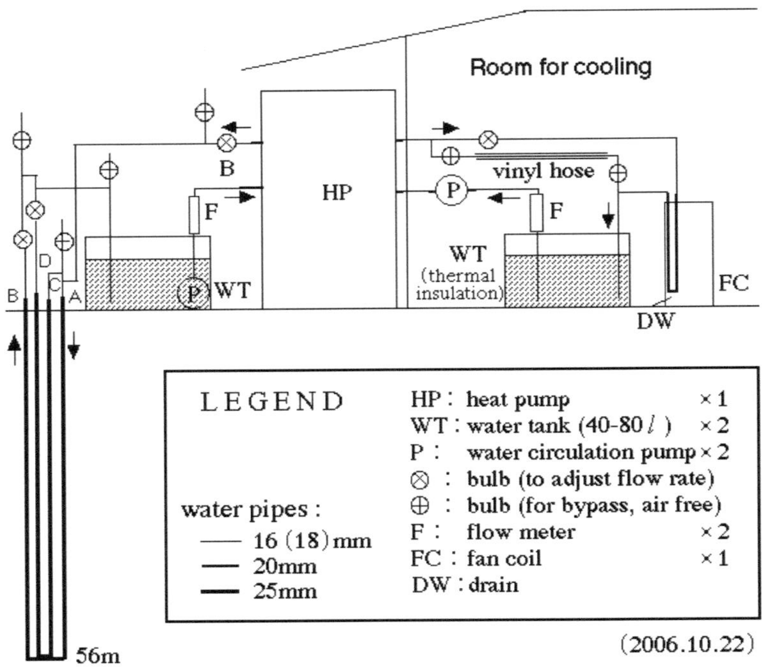

Figure 9.
Schematic figure of the Kamphaengphet GSHP system.

Figure 10.
Outlook of Kamphaengphet GSHP system. Down-left: inside the room, the others: outside.

Figure 11 shows temperature recovery at the bottom of the heat exchange borehole after stopping GSHP operation. During operation of the GSHP system, the subsurface temperature has risen to 33.5°C. It is because the length of the heat exchange pipe is shorter than necessary, and the temperature of circulation water reaches 42°C. However, after stopping system operation (at day 0 of x-axis in **Figure 11**), temperature rapidly drops and recovers to 30.5°C or lower in a week. Even after 15 months of operation in January 2008 (green dashed line), no long-term effect was observed. It may be because of its location near the edge of the basin, where groundwater flow rapidly releases exhaust heat by its advection effect.

In this first experiment, the on-off controlling system of the heat pump was a very basic thermostat, controlled simply by outlet fluid temperature of the heat pump. **Figure 12** shows outlet fluid temperature (light purple, shown as "HP operation temperature"), SCOP (blue-green), outside temperature (green), and room temperature (yellow and blue) during the experiment period. For the first few months (October 19, 2006 to March 7, 2007), the difference between maximum and minimum outlet temperatures was experimentally set to 8 K. However, by this setting, the interval between operation periods of the heat pump was too long, thus the room was not effectively cooled by this system and SCOP was quite low. Then, after trial and error, effective cooling and an SCOP of 3 were achieved with outlet temperature difference of 5 K when outside temperature in daytime was 30–35°C (August 20, 2007 to October 20, 2007).

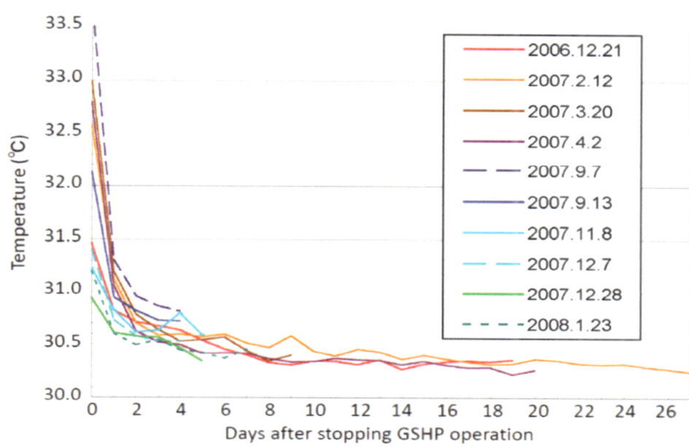

Figure 11.
Temperature recovery at the bottom of heat exchange borehole after stopping GSHP operation [7].

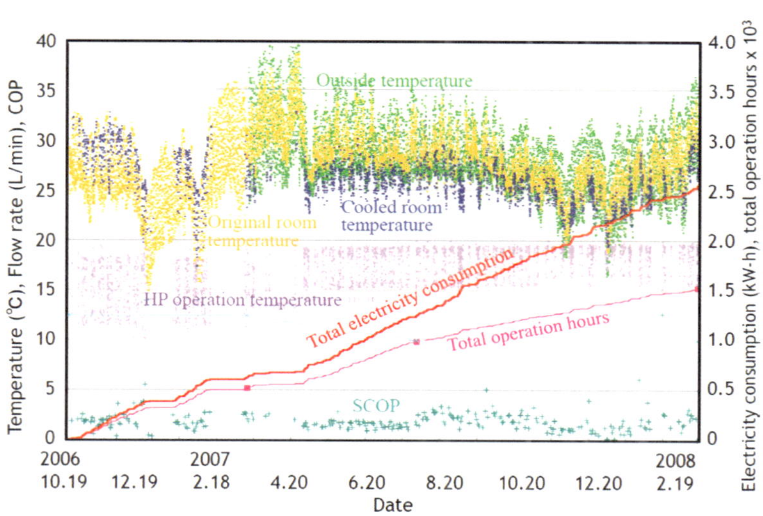

Figure 12.
Temperature monitoring result [7].

It is interesting that atmospheric temperature is not very different from original subsurface temperature, but higher performance is obtained by GSHP as was described by [6]. However, in a season when outside temperature in daytime dropped to 23–32°C (November 19, 2007 to February 19, 2008), SCOP dropped to 2 or lower. Here, SCOP was calculated as follows:

$$\mathrm{SCOP} = \text{provided heat/total electricity consumption} = (T_{outlet} - T_{inlet}) \times Q_2/W_e,$$

where

T_{outlet} is the fan-coil outlet temperature (= heat pump inlet temperature from fan coil);

T_{inlet} is the fan-coil inlet temperature (= heat pump outlet temperature to fan coil);

Q_2 is the flow rate of the secondary fluid (= flow rate from heat pump to fan-coil);

W_e is the total electricity consumption per unit time (= electricity for heat pump and water circulation pump).

In summary, the following results were obtained:

- A continuous operation of the system causes temperature increase in the heat exchange borehole, but it recovers in a week after operation has stopped.
- No long-term subsurface temperature increase occurred even after a year of operation may be because of advection effects of groundwater flow in this region.
- Proper setting of heat pump operation and system design is essential for effective cooling with higher SCOP. Otherwise, its performance may be lower than regular air-conditioner.
- With a proper setting of heat pump operation, SCOP of around 3 was achieved when outside temperature in daytime was 30°C or higher, which is equivalent to original subsurface temperature.

Thus, the applicability of GSHP in tropical regions was confirmed by this experiment. For more effective utilization with better cost performance and SCOP, some adjustment of the system is recommended.

3.3 Demonstration projects in other places

Table 2 shows experiments and demonstrations of GSHP cooling in tropical region conducted by now. Since the applicability has been already confirmed in Kamphaengphet, the following experiments are aiming at higher SCOP and/or better cost performance. Study for better cost performance includes applicability of local material, local technology, and local human resources.

One method to reduce installation cost is the application of horizontal heat exchanger (Nos. 3, 4, and 7 of **Table 2**). The installation costs of simple tubes horizontally buried at a depth of 2 m in Nos. 3 and 4 were lower than that of vertical ones in an order. When burying a tube, it is important not to bend it to keep a high flow rate with a modest water circulation pump. It is also important to make surface piping as short as possible and cover with thermal insulation material to avoid influence of surface heat. In case of No. 7, more sophisticated heat exchangers were used. **Figure 13** shows two types of horizontal heat exchangers, carpet style and

Figure 13.
Installation of horizontal heat exchangers in Saraburi campus, Chulalongkorn University, Thailand (No. 7).

coil style, applied in No. 7. Although a shallow horizontal system may be affected by surface temperature change, its cost performance may be better especially if a large area is available to bury longer heat exchangers. Combination of horizontal and vertical heat exchangers may be a solution if wide area is not available (No. 2). The circulation fluid will be roughly cooled by horizontal pipe and it will be cooled further by the vertical heat exchanger.

Another method to reduce installation cost is to use local material. Nos. 6 and 7 in **Table 2** were planned to use heat pump by local manufacture and No. 7 was successfully operated. For Nos. 5, 8, and 9, electrofusion welding was used to connect the head part of U-tube with local tubes. Since U-tube is not sold in the countries where GSHP is not common yet and importing whole U-tube of 50 m or longer from other country is so costly, electrofusion welding is quite effective to use local material with imported U-tube head.

For higher performance of vertical heat exchanger, well completion without cementing is essential to efficiently raise heat exchange rate per length by the advection effect of groundwater. This method was applied for Nos. 8 and 9; since drilling without cementing was not common in Thailand and Vietnam, technology transfer was needed. In case of No. 8, drilling with normal bentonite mud was done and mud-cake inside the wellbore was washed away after drilling. However, this washing work was quite hard and time-consuming so that the whole process took a week for full drilling crews. Therefore for No. 9, synthetic polymer was applied as drill mud. In this case, drilling and installation of U-tube took only 3 days for an identical system as in No. 8 because polymer mud does not need washing. The local crews were able to conduct polymer drilling without problem using local drilling machine (upper left of **Figure 14**) with an instruction of an expert from Japan.

Nowadays, more sophisticated heat pumps and their controlling systems, which can be used just like normal air-conditioner, are available for GSHP systems. They automatically turn on and off the heat pump for a certain room temperature setting with energy saving. Such systems are used for Nos. 5–9. As result of such a control system and proper design of subsurface heat exchanger, system No. 5 achieved 30% of electricity saving compared to a latest normal air-conditioner [9].

In summary, the following results were obtained through these projects:

- Application of horizontal subsurface heat exchanger is effective to reduce installation cost of heat exchanger. Combination of horizontal and vertical ones is effective to keep higher heat exchange rate and reduce installation cost.

- Shorter piping and thermal insulation of surface pipe is important for effective cooling. Subsurface pipe for horizontal system should not be bended to keep high fluid circulation with a small water pump.

Figure 14.
Installation of a GSHP system in Vietnam Institute of Geology and Mineral Resources (VIGMR), Hanoi, Vietnam (No. 9).

- Application of local material, such as domestic heat pump and domestic tube is effective to reduce total cost. Electrofusion welding is useful to connect the head part of U-tube and local tubes so that import of whole U-tube is not necessary.

- For higher performance and cost reduction of vertical heat exchanger, well completion without cementing is essential. Polymer mud is quite effective for such drilling and local crew may handle it without problem.

- A sophisticated heat pump control system with a proper of heat exchanger achieved 30% of electricity saving compared to a latest normal air-conditioner.

4. Discussion

4.1 Numerical modeling for heat exchange simulation

Higher performance and appropriate design of heat exchange system (length and the number of borehole heat exchangers) may be achieved by numerical simulation prior to GSHP installation. In monsoon Asia, the advection effect of shallow groundwater on heat exchange rate is so dominant that measurement or estimation of groundwater flow rate is essential to perform such a numerical simulation for each region. Literatures describe methods to develop regional potential maps of GSHP installation based on groundwater flow modeling in a plain and simulation of heat exchange rate at any installation location [10–12]. Groundwater temperature surveys or thermal response tests in the region are essential for such modeling.

4.2 Application of open-loop systems

Only closed-loop systems are introduced in these demonstrations, but open-loop systems should also be considered for future application. Normally, open-loop system has higher installation cost because it needs at least two wells, production well and injection well, and its system design is order-made. However, open-loop systems may achieve higher heat exchange rate and lower running cost [13–15]. Therefore, open-loop systems may have higher cost performance if the following conditions are satisfied:

- Water production from a well is allowed in the area (in many urban areas, water production is prohibited by regulations to avoid land subsidence);
- Shallow aquifer is available so that drilling cost may not be high;
- Large system is planned so that high initial cost may be recovered by low running cost in few years; and
- Enough spacing between production well and injection well is available to avoid temperature interference.

In addition, some new ideas of heat exchange systems, so-called "semi-open "systems, which have benefits of both open- and closed-loop systems, are introduced by several authors [16, 17].

4.3 Consideration on subsurface temperature change

Contamination of subsurface temperature by exhaust heat would be a matter of concern. In all cases shown in this chapter, no long-term effects on subsurface temperature at the site were observed. That means exhaust heat was flown away by advection of groundwater and heat concentration was diluted. However, if an intensive installation of GSHP system would be done in the future, it might raise subsurface temperature in some extent and finally the heat may be released to surface water or atmosphere. Note that, however, effect of such heat release on UHI phenomenon should still be lower than that by normal air-conditioner because GSHP may save electricity for cooling, which means the amount of exhaust heat is smaller. Utilization of hot water from heat pump may be a solution to avoid subsurface temperature increase. Also, to avoid local subsurface temperature increase, numerical simulation prior to the installation and temperature monitoring after installation is recommended.

4.4 Heating operation in semi-tropical region

Hanoi, Vietnam, is not a tropical region but semi-tropical. In the winter season in Hanoi, underground temperature is higher than atmospheric temperature and GSHP can be used as a heater. Atmospheric temperature in the winter of Hanoi is not so low that people normally do not use heater. However, since humidity is quite high, heating systems may provide more comfortable life. As a fact, system No. 9 has been used as heater in the winter, and the visitors of this room have been quite impressed by the comfort of the heat by GSHP that they have never experienced. It may be used for drying as well. Thus, GSHP for both cooling and heating might get a new market in semi-tropical regions.

4.5 Application to other regions

GSHP for cooling may be applied for other tropical regions in the world. It is known that GSHP has higher performance than normal air-conditioner even when subsurface temperature is slightly higher than atmospheric ones [6]. In addition, existence of shallow groundwater may give higher performance of GSHP in tropical regions. As shown in **Table 1**, rock's heat conductivity σ, effective heat conductivity σ_e, and apparent heat conductivity σ_a have a relation of $\sigma < \sigma_e < \sigma_a$ in the saturated zone. Therefore, high heat exchange rate in the subsurface heat exchanger may be achieved with the existence of groundwater flow. Collection of long-term operational data is necessary to show the real value of GSHP for both heating and cooling as pointed out by [18].

5. Conclusions

Possibility of GSHP application in tropical Asia is studied based on groundwater temperature survey data in the Chao Phraya plain, Thailand, and in the Red River plain in Vietnam to be compared with atmospheric temperature data. As a result, in most cities in these areas, subsurface temperature is lower than atmospheric temperature in daytime for most months. Thus, it is suggested that shallow underground may be used as a "cold" heat-source at least in daytime. Therefore, experimental operations of GSHP have been conducted in Thailand, Indonesia, and Vietnam to confirm its applicability. In the first experiment in Thailand, SCOP of 3 was achieved. Aiming at higher performance at low cost, several technical improvements were conducted through the experiments. Use of local material, local technology, and local human resources is a key for better cost performance. Well completion without cementing is one of the key technologies to raise the heat exchange rate of a vertical system and polymer drilling is quite useful for its drilling. Horizontal heat exchangers may reduce the installation cost drastically. Combination of horizontal and vertical systems would be effective when enough space for horizontal system is not available. For better designing of GSHP system in monsoon Asia, numerical modeling of the regional groundwater flow and heat exchange simulation at the target location is recommended to know the proper length and depth of the borehole heat exchanger. In the case of semi-tropical region, heating by GSHP may create a new market for comfortable lives. The studies shown here may be applied to other tropical and semi-tropical regions in the world.

Acknowledgements

The temperature measurement in Thailand and Vietnam was conducted by AIST in cooperation with Department of Groundwater Resources (DGR), Thailand, and Department of Geology and Minerals of Vietnam (DGMV), respectively. GSHP experiments were conducted by AIST and Akita University, Japan, in collaboration with DGR, Department of Mineral Resources (DMR), Kasetsart University, and Chulalongkorn University, Vietnam Institute of Geology and Mineral Resources (VIGMR), Bandung Institute of Technology (ITB), and Western Java Energy Mineral Institute. These international collaborations have been done under CCOP Groundwater Water Sub-Project "Development of Renewable Energy for Ground-Source Heat Pump System in CCOP Regions." Special thanks should be given to Prof. Isao Takashima, Akita University, who designed all these systems and installed them with the authors.

Author details

Kasumi Yasukawa* and Youhei Uchida
National Institute of Advanced Industrial Science and Technology, Koriyama, Japan

*Address all correspondence to: kasumi-yasukawa@aist.go.jp

References

[1] Hung T, Uchihama D, Ochi S, Yasuoka Y. Assessment with satellite data of the urban heat island effects in Asian mega cities. International Journal of Applied Earth Observation and Geoinformation. 2005;**8**:34-48

[2] Genchi H. New countermeasures for mitigating heat-island effect—District heat supply systems, underground thermal energy storage, systems using underground as a heat sink. Energy and Resources. 2001;**22**(4):306-310. (in Japanese)

[3] Yasukawa K, Uchida Y, Tenma N, Taguchi Y, Muraoka H, Ishii T, et al. Groundwater temperature survey for geothermal heat pump application in tropical Asia. Bulletin of the Geological Survey of Japan. 2009;**60**(9/10):459-467

[4] Cermak V. Underground temperature and inferred climatic temperature of the past millennium. Palaeogeography, Palaeoclimatology, Palaeoecology. 1971;**10**:1-19

[5] Jessop AM. Thermal Geophysics. Amsterdam: Elsevier; 1990. 306p

[6] Nam Y, Ooka R. Development of water and air multi-source heat pump system using groundwater circulatory wells. In: Proceedings of Renewable Energy 2008; October 2008; Busan, Korea; 2008. O-GT-27

[7] Yasukawa K, Takashima I, Uchida Y, Tenma N, Lorphensri O. Geothermal heat pump application for space cooling in Kamphaengphet, Thailand. Bulletin of the Geological Survey of Japan. 2009;**60**(9/10):491-501

[8] Tenma N, Yasukawa K, Takashima I, Uchida Y, Lorphensri O, Zyvoloski G. Subsurface thermal influence of experimental geothermal heat pump system operation for space cooling in Kamphaengphet, Thailand. Bulletin of the Geological Survey of Japan. 2009;**60**(9/10):503-509

[9] Chokchai S, Chotpantarat S, Takashima I, Uchida Y, Yasukawa K, Charusiri P. On the new ground source heat pump in Thailand: A case study at Chulalongkorn University, Bangkok, Thailand. In: Proceeding of 52nd CCOP Annual Session (Thematic Session "Geoscience for the Society"); Bangkok, Thailand; 2016

[10] Fujii H, Inatomi T, Itoi R, Uchida Y. Development of suitability maps for ground-coupled heat pump systems using groundwater and heat transport models. Geothermics. 2007;**36**(5):459-472

[11] Shrestha G, Uchida Y, Yoshioka M, Fujii H, Ioka S. Assessment of development potential of ground-coupled heat pump system in Tsugaru Plain, Japan. Renewable Energy. 2015;**76**:249-257

[12] Yoshioka M, Shinichi Takakura S, Uchida Y. Estimation of groundwater flow from temperature monitoring in a borehole heat exchanger during a thermal response test. Hydrogeology Journal. 2016;**26**(3):853-867

[13] Aresti L, Christodoulides P, Florides G. A review of the design aspects of ground heat exchangers. Renewable and Sustainable Energy Reviews. 2018;**92**:757-773

[14] Athresh AP, Al-Habaibeh A, Parker K. The design and evaluation of an open loop ground source heat pump operating in an ochre-rich coal mine water environment. International Journal of Coal Geology. 2016;**164**:69-76

[15] Alzoubi MA, Akhtar S, Fong M, Sasmito AP. Characterization of an open-loop seasonal thermal energy

storage system. Energy Procedia. 2017;**142**:3401-3406

[16] Shrestha G, Uchida Y, Kuronuma S, Yamaya M, Katsuragi M, Kaneko S, et al. Performance evaluation of a ground-source heat pump system utilizing a flowing well and estimation of suitable areas for its installation in Aizu Basin, Japan. Hydrogeology Journal. 2017;**25**(5):1437-1450

[17] Farabi-Asl H, Fujii H, Kosukegawa H. Cooling tests, numerical modeling and economic analysis of semi-open loop ground source heat pump system. Geothermics. 2018;**71**:34-45

[18] Song Y, Minder R, Yasukawa K, Katharina Link K. Toward a better way of estimating energy utilization with geothermal heat pumps: A collaborative study under Annex VIII of IEA geothermal. In: Proceedings of World Geothermal Congress 2015; Melbourne, Australia; 2015. 29011